BARRON'S
THE TRUSTED NAME IN TEST PREP

2026

AP® Physics 1
PREMIUM

Kenneth Rideout, M.S., and
Jonathan Wolf, M.A., Ed.M.

AP® is a registered trademark of the College Board, which is not affiliated with Barron's and was not involved in the production of, and does not endorse, this product.

AP® is a registered trademark of the College Board, which is not affiliated with Barron's and was not involved in the production of, and does not endorse, this product.

© Copyright 2025, 2024, 2023, 2022, 2020, 2018, by Kaplan North America, LLC d/b/a Barron's Educational Series

All rights reserved.
No part of this book may be reproduced in any form or by any means without the written permission of the copyright owner.

Published by Kaplan North America, LLC d/b/a Barron's Educational Series
1515 West Cypress Creek Road
Fort Lauderdale, FL 33309
www.barronseduc.com

ISBN: 978-1-5062-9772-9

10 9 8 7 6 5 4 3 2 1

Kaplan North America, LLC d/b/a Barron's Educational Series print books are available at special quantity discounts to use for sales promotions, employee premiums, or educational purposes. For more information or to purchase books, please call the Simon & Schuster special sales department at 866-506-1949.

About the Authors

Ken Rideout has a B.S. in Honors Physics from Purdue University and an M.S. in Physics from Carnegie Mellon University. He has been teaching high school physics for more than twenty years in the Boston area.

Jonathan Wolf has been teaching physics at both the secondary school and college levels for more than thirty-five years, and has also been an adjunct assistant professor of astronomy at Hofstra University. He has published over forty professional papers in the fields of astronomy, physics, and physics education and served for more than ten years as assistant editor for *The Science Teachers Bulletin* published by the Science Teachers Association of New York State (STANYS). In addition to being the author of Barron's *AP Physics B*, he is also the author of Barron's *College Review: Physics* and previously was co-author of Barron's *SAT Subject Test in Physics*. Jonathan Wolf is currently an adjunct professor of physics at Fairleigh Dickinson University.

Table of Contents

How to Use This Book ix
Barron's Essential 5 x

Introduction 1
    Structure and Scope of the AP Physics Exams 1
    Organization of This Book 2
    Study Skills and Tips 3

DIAGNOSTIC TEST

Diagnostic Test 17
    Answer Key 22
    Answers Explained 23

REVIEW AND PRACTICE

1 Vectors 31
    Coordinate Systems and Frames of Reference 31
    Vectors 32
    Addition of Vectors 33
    Subtraction of Vectors 37
    Addition Methods Using the Components of Vectors 37
    Practice Exercises 40

2 Kinematics 45
    Average and Instantaneous Motion 45
    Acceleration 46
    Accelerated Motion Due to Gravity 51
    Graphical Analysis of Motion 52
    Relative Motion 54
    Horizontally Launched Projectiles 55
    Projectiles Launched at an Angle 56
    Uniform Circular Motion 57
    Practice Exercises 61

3 Forces and Newton's Laws of Motion 67
    Forces 67
    Newton's Laws of Motion 69
    Static Applications of Newton's Laws 71
    Free-Body Diagrams 72
    How to Draw Effective Free-Body Diagrams 74
    Dynamic Applications of Newton's Laws 75
    Central Forces 76

    Friction 77
    Practice Exercises 82

4 Energy 91
    Work 91
    Power 94
    Kinetic Energy and the Work-Energy Theorem 94
    Potential Energy and Conservative Forces 96
    Conservation of Energy and Systems 96
    Practice Exercises 102

5 Gravitation 107
    Newton's Law of Universal Gravitation 107
    Inertial and Gravitational Mass 109
    Gravitational Energy 109
    Orbiting Satellites 110
    Practice Exercises 114

6 Impacts and Linear Momentum 119
    Internal and External Forces 119
    Impact Forces and Momentum Changes 119
    The Law of Conservation of Linear Momentum 121
    Elastic Collisions 122
    Inelastic Collisions 123
    Center of Mass 124
    Practice Exercises 127

7 Rotational Motion 133
    Parallel Forces and Moments 133
    Torque 135
    More Static Equilibrium Problems Using Forces and Torques 138
    Rotational Inertia 139
    Angular Kinematics 139
    Energy Considerations for Rolling Objects 141
    Angular Momentum 141
    Practice Exercises 144

8 Oscillatory Motion 149
    Simple Harmonic Motion: A Mass on a Spring 149
    Simple Harmonic Motion: A Simple Pendulum 152
    The Dynamics of Simple Harmonic Motion 154
    Practice Exercises 157

9 Fluids **161**
  Static Fluids 161
  Pascal's Principle 161
  Static Pressure and Depth 162
  Buoyancy and Archimedes' Principle 163
  Fluids in Motion 164
  Bernoulli's Equation 165
  Practice Exercises 170

PRACTICE TESTS

Practice Test 1 .... **179**
  Answers Explained 191
  Test Analysis 200

Practice Test 2 .... **203**
  Answers Explained 215
  Test Analysis 223

APPENDIX

Table of Information for AP Physics 1 .... **227**

Formula Sheet for AP Physics 1 .... **228**

Glossary **229**

Index **235**

How to Use This Book

Diagnostic Test

First, take the diagnostic test to gain an understanding of your strengths and weaknesses. Read the answer explanations for all questions, as they provide valuable insight into correct and incorrect answers. Refer to the Answer Key to identify the areas you need to brush up on. Here you can find the topic and chapter numbers that correspond to each of the questions in the diagnostic test.

Review and Practice

This book's review chapters align with the curriculum for the AP Physics 1 course. You may find it helpful to read the text along with your textbook when you are first learning the material and studying for course exams, or you may choose to read the chapters together as a review after you have completed most of your AP Physics 1 course. By answering the practice questions that follow each chapter, you will be able test your learning as you progress through the book.

Practice Tests

The final section of the book offers the opportunity to take two full-length practice tests that include all question types found on the actual exam. A comprehensive answer explanation is provided for each question.

Online Practice

In addition to the diagnostic test and two practice tests within this book, there are also two full-length online practice exams. You can take these exams in practice (untimed) mode or in timed mode. All questions include answer explanations.

BARRON'S ESSENTIAL 5

As you review the content in this book to work toward earning that **5** on your AP Physics 1 exam, here are five things that you **MUST** know above everything else:

1 **Know your kinematics:**
- Know the difference between velocity and speed, displacement and distance.
- Know to use equations of motion only for problems involving constant acceleration and to otherwise use graphical methods to analyze the kinematics.
- Find hidden information within the problem (initial or final speeds of zero, accelerations of 9.8 m/s^2).
- Keep your *x*- and *y*-motions separate: the only connecting variable is time.

2 **Know your dynamics:**
- Start all analyses with a free-body diagram.
- Align your coordinate system with the direction of acceleration (if known).
- Be alert for situations that require a net force but no corresponding change in speed (centripetal forces).
- Do not add additional forces in an ad hoc manner. Decide on the number of forces based on the object's interactions, not its motion.

3 **Know your conservation laws:**
- If all the masses and motions of interacting particles are specified, conserve the net values of energy, momentum, and angular momentum.
- If the object in question is interacting with an object whose masses and motions are unspecified, then use the interactions to calculate changes in "conserved" quantities.
- When using conservation of energy, be alert to work done by nonconservative forces.
- When using conservation of momenta, remember that both linear and angular momenta are vector quantities.

4. Look for cross-cutting questions:

- This exam is designed to probe for your understanding of the connections between the various topics covered in a first-year physics class. It is not enough to be able to solve the "classic" problem types. You must be able to solve multitiered problems (e.g., apply conservation of angular momentum and dynamics at the same time within one problem) and apply your knowledge to new situations.

5. Understand the underlying concepts:

- Be sure to understand how to explain the why's, not just to calculate a numerical solution. Novel situations or new types of problems can be solved by students who understand the concepts behind the equations.

Introduction

Learning Objectives

In this chapter, you will learn about:

- → Units
- → Relationships and review of mathematics
- → Tips for answering multiple-choice questions
- → Tips for solving free-response questions
- → Graphs, fits, and the linearization of data
- → Uncertainty and percent error
- → Study skills and scheduling your review
- → Objects versus systems

Structure and Scope of the AP Physics Exams

The College Board currently offers four AP Physics exams. This book is for students preparing for the AP Physics 1 exam, which corresponds to a first-year algebra-based college course. The other exams are AP Physics 2 (a second-year algebra-based college course), AP Physics C Mechanics (calculus based), and AP Physics C Electricity and Magnetism (also calculus based). Starting in May 2025, standard paper testing will be discontinued for this exam. You will complete multiple-choice questions and view free-response questions in the Bluebook app. You will handwrite your free-response answers in paper exam booklets that are returned for scoring. Both the AP Physics 1 and AP Physics 2 exams focus on conceptual underpinnings and basic scientific reasoning along with the traditional problem-solving aspects of physics. In addition, both exams have questions that require experiential lab understanding. Although there are some calculation-oriented questions, these two tests are explicitly designed not to be "plug and chug" type questions. Students who do not thoroughly understand the physics concepts behind the equations will find themselves at a disadvantage.

> **AP Physics 1 Test Basics**
> - 80 minutes for 40 multiple-choice questions
> - 100 minutes for 4 free-response questions
> - Calculator allowed throughout
> - Formula sheet and table of information provided throughout

The AP Physics 1 exam focuses on mechanics (including rotational mechanics), the three conservation laws (energy, linear momentum, and angular momentum), fluids, and oscillations. The AP Physics 2 exam picks up many of the other topics typically found in an introductory sequence in college physics and also covers some of the overlapping areas, such as force fields, electricity, and waves. Formulas are provided for you during the test (see the appendix). However, it is important that you not only know where and what the provided equations mean, but you must also be able to determine quickly under what situations the equations can and should be used. Even if a question is conceptual, having a corresponding equation in mind can guide your thinking. Note that some useful equations are not given (e.g., orbital velocity, Kepler's third law, rotational inertia, projectile motion) and that some given equations (universal gravitational potential energy, for example) are unlikely to be used. The equation sheet does, however, provide a solid foundation. A well-prepared student will be able to find and understand every equation on it.

Although you may not use your calculator often during the exam, one is allowed throughout. Check the College Board website for an approved list of calculators. Generally, all scientific and graphing calculators are allowed. You should make sure you have extra batteries for your calculator during the exam. A ruler is also permitted, but its usefulness is likely limited to drawing straight lines, if needed, during the free-response section.

Organization of This Book

This introduction serves as general background to the AP exam. A diagnostic test follows. This exam is intended as a tool to help already-prepared students determine if they have any weakness in content and to direct them to the appropriate chapter in this book for review. Although example problems and targeted questions are included at the end of each chapter to reinforce the content, the full-length practice tests are where you will find questions that are most closely modeled after the actual AP Physics 1 exam. Each chapter covers a specific topic in physics, so the questions in each chapter are limited to that specific concept. The specific problem types in the AP Physics 1 exam are mirrored in the practice tests.

Study Skills and Tips

Units

Preparing for an AP exam takes time and planning. In fact, your preparation should begin in August or September when you start the class. If you are using this review book during the year, the content review chapters could parallel what you are covering in class. If you are using this review book a few weeks prior to the exam in May, your strategy needs to change. The review material should help you refresh your memory as you work on the practice tests. In either case, you should have a plan.

In this chapter, we will look at study skills and tips for helping you do well on the AP Physics 1 exam. One of the most important things to remember is that most physical quantities have units associated with them. You must memorize units since you may be asked questions about them in the multiple-choice section. In the free-response questions, you must include all units when using equations, making substitutions, and writing final answers.

A list of standard fundamental (SI) units as well as a list of some derived units are shown in the following two tables. As you work through the different chapters, make a note (for example, on index cards) of each unit.

TIP
Make sure you set up a review schedule.

TIP
Make sure you memorize all units. Be sure to include them with all calculations and final answers.

Fundamental SI Units Used in Physics

Quantity	Unit Name	Abbreviation
Length	Meter	m
Mass	Kilogram	kg
Time	Second	s
Electric current*	Ampere	A
Temperature*	Kelvin	K
Amount of substance*	Mole	mol
Luminous intensity*	Candela	cd

*Not likely to appear on the AP Physics 1 exam

Some Derived SI Units Used in Physics

Quantity	Unit Name	Abbreviation	Expression in Other SI Units
Area			m^2
Linear velocity			m/s
Linear acceleration			m/s^2
Force	Newton	N	$kg \cdot m/s^2$
Momentum			$kg \cdot m/s$
Impulse			$N \cdot s = kg \cdot m/s$
Angular velocity			rad/s
Angular acceleration			rad/s^2
Torque			$N \cdot m$
Angular momentum			$kg \cdot m^2/s$
Rotational inertia			$kg \cdot m^2$
Spring constant		N/m	kg/s^2
Frequency	Hertz	Hz	s^{-1}
Pressure	Pascal	Pa	$N/m^2 = kg/(m \cdot s^2)$
Work, energy	Joule	J	$N \cdot m = kg \cdot m^2/s^2$
Power	Watt	W	$J/s = kg \cdot m^2/s^3$

Relationships and Review of Mathematics

Since AP Physics 1 is an algebra-based course, the appendix reviews some essential aspects of algebra. In physics, we often discuss how quantities vary using proportional relationships. Four special relationships are commonly used. You can review them in more detail by referring to the appendix. You should memorize these relationships.

- **Direct relationship**—This is usually represented by the algebraic formula $y = kx$, where k is a constant. This is the equation of a straight line, starting from the origin. An example of this relationship is Newton's second law of motion, $\vec{a} = \frac{\vec{F}_{net}}{m}$, which states that the acceleration of a body is directly proportional to the net force applied (see Chapter 3).

  > **Reminder**
  > These relationships are also useful for analyzing data to answer laboratory-based questions. A laboratory-based question is always on the exam.

- **Inverse relationship**—This is usually represented by the algebraic formula $y = \frac{k}{x}$. This is the equation of a hyperbola. An example of this relationship can be seen in a different version of Newton's second law, $\vec{F}_{net} = m\vec{a}$. In this version, if a constant net force is applied to a body, the mass and acceleration are inversely proportional to each other. Some special relationships, such as gravitation and static electrical forces, are known as inverse square law relationships. The forces are inversely proportional to the square of the distances between the two bodies (see Chapters 5 and 9).

- **Squared (quadratic) relationship**—This is usually represented by the algebraic formula $y = kx^2$ and is the equation of a parabola starting from the origin. An example of this relationship can be seen in the relationship between the displacement and uniform acceleration of a mass from rest $\vec{d} = \frac{1}{2}\vec{a}\,t^2$ (see Chapter 2).
- **Square root relationship**—This is usually represented by the algebraic formula $y = k\sqrt{x}$ and is the equation of a "sideways" parabola. This relationship can be seen in the relationship between the period of a simple pendulum and its length, $T = 2\pi\sqrt{L/g}$ (see Chapter 8).
- As you review your material, you should know each of these relationships and their associated graphs (see the appendix for more details).

Tips for Answering Multiple-Choice Questions

Without a doubt, multiple-choice questions can be tricky. The AP Physics 1 exam contains 40 multiple-choice questions. These can range from a simple recall of information to questions about units, graphs, proportional relationships, formula manipulations, and simple calculations.

One tip to remember is that there is no penalty for wrong answers. This means that you want to try to answer all questions. Instead of randomly guessing, however, you can improve your chances of getting a correct answer if you can eliminate at least two answer choices. Guess intelligently. For the Physics 1 exam, all multiple-choice questions will have four answer choices.

When you read a multiple-choice question, try to get to the essential aspects. You have 80 minutes for this part, so do not waste too much time per question. Try to eliminate two or three choices. If a formula is needed, you may try to use approximations (or simple multiplication and division).

According to the College Board, the breakdown by topic of the multiple-choice questions are as follows.

Kinematics	10–15%
Force and Translational Dynamics	18–23%
Work, Power, and Energy	18–23%
Linear Momentum	10–15%
Torque and Rotational Dynamics	10–15%
Rotating Systems (energy and momentum)	5–8%
Oscillations	5–8%
Fluids	10–15%

As you work on the multiple-choice questions in the practice tests, look for distractors. These are choices that may look reasonable but are incorrect. For example, if the question is expecting you to divide to get an answer, the distractor may be an answer obtained by multiplying. Watch out for quadratics (such as centripetal force) or inverse squares (such as gravitation), where linear reasoning does not apply.

> The magnitude of the acceleration due to gravity (g) can be approximated as 10 m/s². You can also use estimations or order of magnitude approximations to see if answers make sense.

If you cannot recall some information, perhaps another similar question will cue you as to what you need to know. (You may work on only one part of the exam at a time.) When you read the question, try to link it to the overall general topic, such as kinematics, dynamics, or electricity, and then narrow down the specific area and the associated formula. Finally, you must know which quantities are vectors and which quantities are scalars (see Chapter 1).

Each multiple-choice question in the practice tests is cross-indexed with the general topic area of physics to guide you on your review. As you work on the tests and check your answers, you can easily go back to the topic area to review. At the start of your review, you may want to work on the multiple-choice questions untimed for the diagnostic and first practice tests. A few days before the AP exam (see the timeline schedule later in this chapter), you should do the last practice test timed (80 minutes for the multiple-choice section and an additional 100 minutes for the free-response section).

Tips for Solving Free-Response Questions

The AP Physics 1 exam includes four free-response questions, in the following order:

Order	Topic	Suggested Time	Points
1	Mathematical Routines	20–25 minutes	10
2	Translation Between Representations	25–30 minutes	12
3	Experimental Design and Analysis	25–30 minutes	10
4	Qualitative/Quantitative Translation	15–20 minutes	8

The mathematical routines question will have you use mathematics to analyze a scenario and make predictions about that scenario by symbolically deriving relationships and calculating numerical values.

The translation between representations question will have you connect different representations of a scenario. You will draw graphs that relate quantities within the scenario. Some of these quantities will need to be mathematically derived. You may also be asked to do any of the following: justify why answers to any two of the previous parts do or do not agree, make a prediction about another situation and justify it based on your graph, and make a prediction (and justify it!) about how those representations would change if properties of the scenario were altered.

The experimental design and analysis question will have you create scientific procedures in roughly two parts:

1. For design and analysis, your experiment should vary only a single parameter and should measure how that change affects a single characteristic. Describe only methods and equipment that would make sense in a typical high school lab. Describe how the collected data could be analyzed in order to answer the posed question.
2. You will then be given experimental data and answer a similar, but not identical, question. Use the data provided to create a graph that can be analyzed to determine the answer to the given question. Slope or intercepts of your line of best fit may be critical to answer the question.

The qualitative/quantitative translation question will have you connect the nature of the scenario, the physical laws that govern the scenario, and mathematical representations of that scenario to each other. You will then justify a claim about a given scenario and derive an equation related to that scenario. Also, you will be asked to do one of the following: justify why answers to any two of the previous parts do or do not agree, make a prediction about another situation and justify it, and make a prediction (and justify it!) about how those representations would change if properties of the scenario were altered.

You will have 100 minutes for this section. Each question will have specific point allocations for each subsection. However, the free-response portion of the exam in its entirety is worth 50 percent of your "raw score." Keep in mind that the curve determining your actual reported score (out of 5), based on the raw score from the points you earned, changes from year to year.

According to the College Board, their three science practices are weighted in the free-response section as follows.

Scientific Practice	Expectations	Weighting in Free-Response Section
Creating Representations	Diagrams, tables, or schematics. Quantitative graphs (scales, units, labels!). Qualitative sketches of graphs capturing essential features.	20–35%
Mathematical Routines	Deriving symbolic expressions. Calculating or estimating an unknown. Comparing two or more situations. Predicting changes in values based on functional dependence of quantities.	30–40%
Scientific Questioning and Argumentation	Creating an experimental procedure. Application of a law or relationship to make a claim. Justifying a claim using evidence.	35–45%

Since you are not given specific formulas for some concepts, you should begin learning how these formulas are derived starting at the beginning of the year. For example, you are not given the specific formulas for projectile motion problems since these are easily derived from the standard kinematics equations. If you begin reviewing a few weeks before the AP exam, you may want to make index cards of formulas to help you to memorize them.

You must read the entire question carefully before you begin. As you begin to solve the problem, make sure that you write down the general concept being used—for example, conservation of mechanical energy or conservation of energy. Then, you must write down the equations you are using. For example, if the problem requires you to use conservation of mechanical energy (potential and kinetic energies), write out those equations:

Initial total mechanical energy = Final total mechanical energy

$$mgh_i + \frac{1}{2}mv_i^2 = mgh_f + \frac{1}{2}mv_f^2$$

When you are making substitutions, you should indicate to the grader what you are doing. For example, if you are calculating the net force on a mass, you should write as neatly as possible:

$$\sum F = F_{net} = ma = (2\text{ kg})(4\text{ m/s}^2) = 8\text{ N}$$

Include all relevant information. Communicate with the grader by showing him/her that you understand what the question is asking. You may want to make a few sketches or write down your thoughts in an attempt to find the correct solution path. When a written response is requested, make sure that you write neatly and answer the question in full sentences.

Sometimes the question refers to a lab experiment typically performed in class or to simulated data that are given. In that case, you may be asked to make a graph. Make sure the graph is labeled correctly (with axes labeled and units clearly marked), points plotted as accurately as possible, and best-fit lines or curves used. Do not connect the dots. Always use the best-fit line for calculating slopes. Make sure you include your units when calculating slopes. Always show all of your work.

If you are drawing vectors, make sure the arrowheads are clearly visible. For angles, there is some room for variation.

Since angles are measured in degrees, be sure your calculator is in the correct mode. If scientific notation is used, make sure you know how to input the numbers into your calculator correctly. Remember, each calculator is different.

> **TIP**
>
> Make sure you show all of your work on Part II. Include all formulas, substitutions with units, and general concepts used. Remember to label all diagrams. Communicate with the grader!

If you are asked to draw a free-body diagram (see pages 74–75), make sure you include only actual applied forces. Do not include component forces. Centripetal force is not an applied force and should not be included on a free-body diagram.

What do you do if you are not sure how to solve a problem? Follow these 11 tips.

1. Make sure you understand the general concepts involved and write them down.
2. Write down all appropriate equations.
3. Try to see how this problem may be similar to one you may have solved before.
4. Make sure you know which information is relevant and which information is irrelevant to what is being asked.
5. Rephrase the question in your mind. Maybe the question is worded in a way that is different from what you are used to.
6. Draw a sketch of the situation if one is not provided.
7. Write out what you think is the best way to solve the problem. This sometimes triggers or cues a solution.
8. Use numbers or estimations if the solution is strictly algebraic manipulation, such as deriving a formula in terms of given quantities or constants.
9. Relax. Sometimes if you move on to another problem, take a deep breath, close your eyes, and relax for a moment, the tension and anxiety may go away and allow you to continue.
10. Do not leave anything out. Unlike on the multiple-choice questions, you need to show all of your work to earn credit. If you cannot solve the first part of a problem, you can always declare an answer and use it for full credit on the subsequent sections.
11. Understand what you are being asked to do. The Physics 1 exam wants you to respond in specific ways to certain key words.

TIP

Make sure you have pencils, a calculator, extra batteries, and a metric ruler with you for the exam!

Task	The College Board Expects ...
"Calculate"	Provide numerical and algebraic work leading to the final answer. Don't forget to include units and significant figures!
"Compare"	Elaborate on similarities and/or differences.
"Derive"	Starting with a fundamental equation (such as those given on the formula sheet), mathematically manipulate it to the desired form.
"Describe"	List the relevant characteristics.
"Determine"	After explaining or calculating, arrive at a conclusion.
"Draw"	Create a diagram showing physical objects and their relationships.
"Estimate"	Roughly calculate (to the closest power of 10), indicate greater than or lesser than, or indicate positive/negative values. No need to show work.
"Indicate"	Simply provide information (without explanation).
"Justify"	Provide qualitative reasons (not mathematical) to support your claim.
"Label"	Indicate unit, scale, or components in a graph or other representation.
"Plot"	Place specific data points onto a scaled grid. Do not connect the dots (although trends, especially linear ones, may be superimposed on the graph).
"Rank"	Order by magnitude.
"Sketch"	Without numerical scaling or specific data points, draw a representation that captures the key trend in the relationship (curvature, asymptotes, and so on).
"Verify"	Show that the specific condition is met, and explain why it applies.

Graphs, Fits, and the Linearization of Data

Graphs that are linear in nature are much easier to analyze, especially by hand, than graphs of any other nature. Trends, slopes, intercepts, and correlations of experimental results to theoretical predications are readily obtained. For this reason, if you are asked to graph your data, it will almost always be advantageous to linearize it first. Specifically, if the relationship is not linear to start, use a change of variable to make the relationship linear.

For example, if asked to determine the spring constant k of a system based on a collection of elastic potential energies for various extensions of the spring, the relevant equation is

$$U_s = \frac{1}{2}kx^2$$

This is a quadratic relationship, not a linear one. Graphing energy versus displacement will therefore produce a parabola shape from which it is difficult to extract the spring constant k. A better choice is to linearize the data before graphing. Before graphing, make the following change of variable:

$$z = x^2$$

So the relationship is now:

$$U_s = \frac{1}{2}kz$$

Now when graphed and a line of best fit is applied, the slope of the straight line will be ½k.

How can a line of best fit be generated by hand? If need be, you can use a straightedge and draw one straight line that has as many data points above the line as below it. Once this line is drawn, all subsequent calculations should be based on the slope and intercept of this best-line fit rather than on the original data. The idea here is that the fit of the data is an average of the raw data and is inherently better than any one particular point because the random variations in data have been smoothed out by the fit. When asked to analyze a graph of data, always use the fitted line or curve rather than the individual data points for this same reason.

> **Sample Problem**

Determine k by graphing the following data provided by another student.

U_s (J)	x (cm)
0.058	2.5
0.196	4.6
1.117	11.0
2.081	15.1

✓ Solution

Since the relationship between these variables is quadratic, begin by squaring the given values for x. Also change to the standard MKS units of meters. Note that you must scale the axis yourself based on the given data. Your scale should make sure that at least two-thirds of the graph space provided is used.

x (cm)	x^2 (m²)
2.5	0.000625
4.6	0.002116
11.0	0.0121
15.1	0.022801

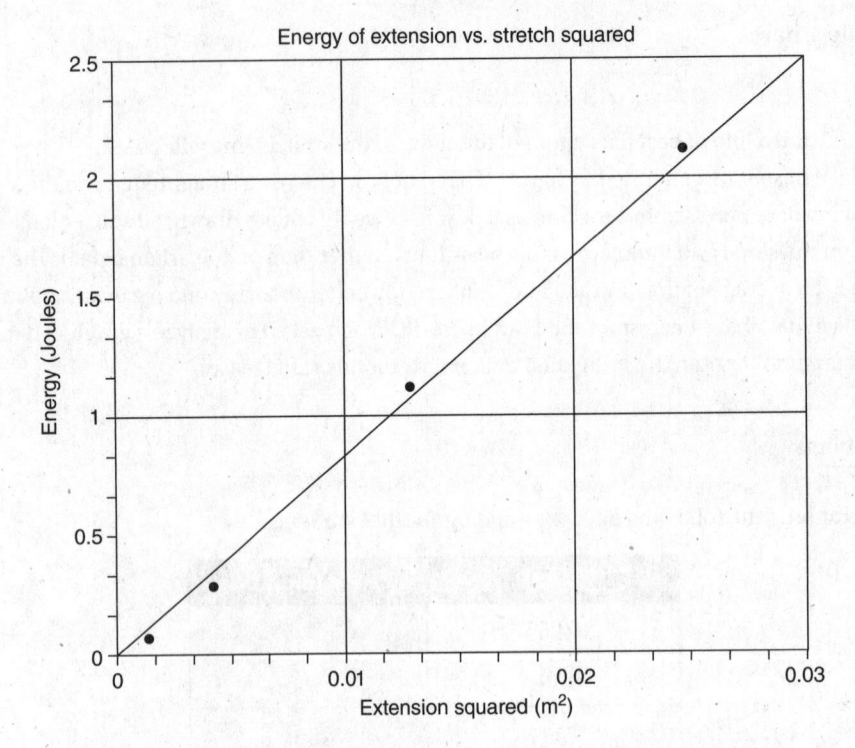

Energy of extension vs. stretch squared

Approximate slope = $\Delta y/\Delta x$ = 2.5/0.03 = 83 J/m² = 83 N/m

Note the labels (with units) on the x- and y-axes as well as the title for the graph.

Uncertainty and Percent Error

In addition to respecting the number of significant digits in measured or recorded data, you can also determine the corresponding uncertainty in derived quantities. For example, if the radius of a circle is measured and recorded as 3.5 cm, there are only 2 significant digits in this number. Therefore, the area of the circle (πr^2) should be truncated from the calculator result to 2 significant digits. The area is 38 cm². The rest of the digits are not significant as they imply a precision in the radius that we do not have.

To take this analysis one step further, a percent error can be associated with a measurement. For example, one could write down the uncertainty in radius explicitly as $r = 4.5 +/- 0.05$ cm. This implies that the true value is most likely between 4.45 and 4.55 cm. This is a percent error of 1.1%.

$$\frac{0.05}{4.5} \times 100 = 1.1\%$$

$$100 * \text{Uncertainty/Value} = \text{Percent error}$$

Percent errors are an easier way of comparing the relative precision of different measurements. For example, a measurement of $128 +/- 2$ mm has a percent error of 1.6% and is thus less precise than our original measurement that has an error of 1.1%.

The College Board offers the following clarification about their expectations on uncertainty: "On the AP Physics 1 exam, students will not need to calculate uncertainty but will need to demonstrate understanding of the principles of uncertainty." Although a student should be able to use, reason, and support his or her answers by proper use of significant digits and percentage error, the student will not be expected to propagate errors, calculate standard deviations, or carry out formal linear regressions.

Study Skills and Scheduling Your Review

Preparing for any Advanced Placement exam takes practice and time. Effective studying involves managing your time so that you efficiently review the material. Do not cram a few days before the exam. Getting a good night's sleep before the exam and having a good breakfast the day of the exam is a better use of your time than "pulling an all-nighter." Working in a study group is a good idea. Using index cards to make your own flash cards of key concepts, units, and formulas can also be helpful.

When you study, try to work in a well-lighted, quiet environment, when you are well rested. Studying late at night when you are exhausted is not an effective use of your time. Although some memorization may be necessary, physics is best learned (and studied) by actively solving problems. Remember, if you are using this book during the year, working through the chapter problems as you cover each topic in class, memorizing the units, and familiarizing yourself with the formulas at that time will make your studying easier in the days before the exam.

If you are using this book in the weeks before the exam, make sure you are already familiar with most (if not all) of the units, equations, and topics to be covered. You can either use the chapter review for a quick overview and practice, or dive right in to the diagnostic test. You do not need to take the diagnostic test under timed conditions. See how you do, and then review the concepts for those questions that you got wrong. You can use the end-of-chapter questions to test your grasp of specific topics and then work on the remaining practice tests.

Setting up a workable study schedule is also vital to success. Each person's needs are different. The following schedule is just one example of an effective plan.

> **Conventions**
>
> The following conventions are used on the AP Physics 1 exam and in this book:
>
> - The frame of reference of any problem is assumed to be inertial (unless otherwise stated).
> - Assume air resistance is negligible (unless otherwise stated).
> - In all situations, positive work is defined as work done on a system.

> Don't forget that you have *two* full-length practice tests with this book, in addition to the diagnostic test.

Objects Versus Systems

An **object** is thought of as an isolated mass that is being acted upon by outside forces. Any internal structure of the object is ignored. The object is defined by its properties (e.g., mass, charge). Strictly speaking, for a single object being analyzed, it does not make sense to make use of potential energy, Newton's third law, thermal energy, or the conserved quantities. Instead, when discussing a single object, use the following relationships from mechanics:

$$\vec{F}_{net} = m\vec{a}$$
$$\vec{\tau} = I\vec{\alpha}$$
$$\vec{F}\Delta t = m\Delta \vec{v}$$
$$F\Delta d = \Delta KE$$

A **system**, in contrast, is made up of objects that may interact. It is within isolated systems that the conserved quantities, canceling Newton's third law forces (internal forces), the heat capacity of the constituent parts, and the potential energy of relationship between interacting objects are useful concepts. Within systems, conserved quantities become the most useful lens through which to view the situation:

$$\sum \vec{p}_i = \sum \vec{p}_f$$
$$\sum \vec{L}_i = \sum \vec{L}_f$$
$$ME_i = ME_f$$

Test Prep Schedule

September 1–April 15	As the year progresses, make sure you memorize units and are comfortable with formulas. If you are using this book during the year, do end-of-chapter problems as they are covered in class. Make sure you register for the exam, following school procedures, and refer to the College Board's website for details: www.collegeboard.com
Four weeks before the exam	Most topics should be covered by now in class. If you are using this book for the first time, begin reviewing concepts and doing the end-of-chapter problems. Begin reviewing units and formulas. Devote at least 30 minutes each day to studying.
Three weeks before the exam	Start working on the diagnostic test. Go back and review topics that you are unsure of or feel that you answered incorrectly.
Two weeks before the exam	Begin working on practice tests. Continue to review old concepts.
One week before the exam	Do the remaining practice tests timed. Make sure you are comfortable with the exam format and know what to expect. Review any remaining topics and units.
The day before the exam	Pack up your registration materials, pencils, calculator, extra batteries, and metric ruler. Put them by the door, ready to go. Get a good night's sleep.
The day of the exam	Have a good breakfast. Make sure you take all the items you prepared the night before. Relax!

SUMMARY

- Make sure you set up a manageable study schedule well in advance of the exam.
- Make sure you memorize all units and are familiar with the exam format.
- Multiple-choice questions do not have a penalty for wrong answers, so do not skip any. If you are unsure of the answer, try to eliminate as many choices as you can, and then guess!
- Do not leave any question out on the free-response part! Show all of your work. Write down all fundamental concepts, write all equations used, and include units for all substitutions and in your final answer.
- Read each question carefully. Write your answers clearly. Write out short-answer questions in full sentences. Clearly label graphs with units and use best-fit lines or curves.
- An object can be represented by a single mass. When exposed to external forces, it is best modeled as experiencing changes in speed and/or direction.
- A system is a group of objects which, if isolated, is best modeled with conservation laws.
- Try to relax and do all of the practice tests. Work on the chapter questions to review concepts as needed.
- Get a good night's sleep before the exam.
- On the day of the exam, bring all registration materials with you, as well as pencils, calculators, extra batteries, and a metric ruler.

Relax and Good Luck!

Diagnostic Test

This section contains a short diagnostic test. The purpose of this diagnostic exam is for you to identify those conceptual areas most in need of review. The relevant sections of the book (indicated in the Answer Key) should be reviewed thoroughly before attempting one of the full-length practice tests at the end of this book.

Diagnostic Test

AP Physics 1
Section I: Multiple-Choice

DIRECTIONS: Select the best answer on each of the following 21 multiple-choice questions as well as the 2 free-response questions. The detailed answer explanations will direct you to a specific chapter for further review on the specific subject matter in that question. You may use a calculator and make use of the formula sheet provided in the appendix.

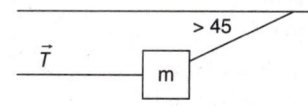

1. A mass is suspended from the ceiling by one rope at an angle as indicated. The mass is held in place by a second, horizontal rope. Without knowing the exact angle from the ceiling, except that it is greater than 45 degrees, which statement best represents the possible value for tension in the horizontal line?

    (A) $T > mg$
    (B) $T = mg$
    (C) $T < mg$
    (D) Cannot be determined without knowing the angle

2. Pushing on a mass of 15 kg with a sideways force of 120 N is not enough to get the mass to start sliding across the floor. Which of the following best describes the coefficient of static friction between the floor and the mass?

    (A) $\mu = 0.8$
    (B) $\mu > 0.8$
    (C) $\mu < 0.8$
    (D) $\mu = 1.25$

3. While standing on an elevator and experiencing a downward acceleration of 4.9 m/s², what is the force between the floor and your feet? The variable m is your mass.

    (A) mg
    (B) $0.5mg$
    (C) $1.5mg$
    (D) $4.9mg$

4. After throwing a ball at an upward angle, the reason the ball continues to go horizontally while falling due to gravity is that

    (A) the ball's inertia keeps it going
    (B) the force of the throw keeps it going
    (C) the energy from the gravitational potential keeps replenishing the lost energy
    (D) air pressure keeps the ball from falling too quickly

5. Consider the situation in which your hand is pushing a book across a rough desktop with lots of friction. Of the many forces involved, consider only these two individual forces: the force on your hand from the book and the force the book is experiencing from your hand. While the book is accelerating from rest to its final velocity, which statement best compares the force experienced by your hand compared with that experienced by the book?

   (A) Force on book > force on hand
   (B) Force on hand > force on book
   (C) Force on hand = force on book
   (D) The relationship between these two forces depends on the frictional force.

6. A puck comes to a stop on a level floor in 20 meters after an initial speed of 10 m/s. What is the coefficient of kinetic friction between the puck and the floor?

   (A) The mass is required to calculate this answer.
   (B) 0.25
   (C) 0.5
   (D) 0.025

7. A rock is dropped from an extremely high cliff and experiences free-fall conditions. Which of the following statements is NOT true about the rock's velocity, acceleration, or displacement between the 3rd and 4th seconds of falling?

   (A) The rock will fall an additional 4.9 meters.
   (B) The rock will speed up by 9.8 m/s.
   (C) The rock's acceleration will remain 9.8 m/s².
   (D) The rock's displacement, velocity, and acceleration vectors are all directed downward.

8. When using the kinematics equation $x = x_0 + v_{x0}t + \frac{1}{2}a_x t^2$, which of the following is always assumed to be true?

   (A) a_x = zero
   (B) $x > x_0$
   (C) V_{x0} is positive.
   (D) a_x is constant.

9. Consider a projectile launched at 30 degrees above a flat surface. The projectile experiences no air resistance and lands at the same height from which it was launched. Compare the initial and final conditions of the following vector components:

   $$V_x \quad V_y \quad a_x \quad a_y$$

   (A) They are all the same at both instances.
   (B) They are all different from their initial values.
   (C) V_x and V_y have changed, but a_x and a_y are constant.
   (D) They are all the same except for V_y.

10. A physics teacher is swirling a bucket around in a vertical circle. If the bucket slips out of his hand when it is directly overhead, the bucket will

    (A) fly upward initially
    (B) drop straight down
    (C) fly out horizontally initially
    (D) fly out up and at an angle initially

11. If the Moon was twice as far from the center of Earth as it is currently, how would its orbital period change?

    (A) The Moon's orbital period would double.
    (B) The Moon's orbital period would increase by a factor of 4.
    (C) The Moon's orbital period would increase by a factor of $2\sqrt{2}$.
    (D) The Moon's orbital period would decrease by a factor of 2.

12. If you went to another planet, which of the following would be true?

    (A) Your mass would be the same, but your weight would change.
    (B) Your mass would change, but your weight would be the same.
    (C) Your mass would be the same, and your weight would be the same.
    (D) Your mass would change, and your weight would change.

13. You lift a heavy suitcase twice. Each time you lift it to the same height, but the second time you do it in half the time. Compare the work done to and power delivered to the suitcase.

    (A) Same work, same power
    (B) Twice the work, twice the power
    (C) Twice the work, same power
    (D) Same work, twice the power

14. A constant horizontal force of 12 N is applied for 5 meters to move an initially stationary mass of 5 kg. The friction between the floor and the mass does a total of −40 joules of work to the mass. The final speed (in m/s) of the object is

    (A) 2(6)^½
    (B) 2(2)^½
    (C) 2(10)^½
    (D) 4

15. A skier begins a downhill run at a height of 9 meters and a speed of 2 m/s. If you ignore friction, what will be her speed when she is 3 meters from the bottom of the slope?

    (A) 6.0 m/s
    (B) 12.8 m/s
    (C) 11.1 m/s
    (D) 7.7 m/s

16. A rubber ball ($m = 0.250$ kg) strikes a wall horizontally at 3.50 m/s and rebounds elastically. What is the magnitude of impulse delivered to the wall in N · s?

    (A) 0.875
    (B) 1.75
    (C) 2.50
    (D) 8.75

17. A running back ($m = 85$ kg) running at 1.5 m/s is tackled from the side by another player ($m = 75$ kg) running perpendicularly to the running back's original heading at 1.75 m/s. What is the resulting speed of the two entangled players just after the tackle?

    (A) 2.2 m/s
    (B) 0.25 m/s
    (C) 1.6 m/s
    (D) 1.1 m/s

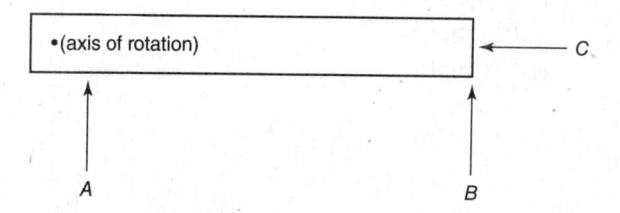

18. Which of the following ranks the torques applied by the three equal forces A, B, and C to the long thin rod about its fixed axis as indicated above?

    (A) $\tau_C > \tau_B > \tau_A$
    (B) $\tau_A = \tau_B > \tau_C$
    (C) $\tau_B = \tau_C > \tau_A$
    (D) $\tau_B > \tau_A > \tau_C$

19. A large spherical cloud of dust in deep space that is spinning slowly collapses under its own gravitational force into a spherical cloud 10 times smaller in diameter. How does its rotational speed change?

    (A) The rotational speed does not change.
    (B) The rotational speed increases by a factor of 10.
    (C) The rotational speed decreases by a factor of 10.
    (D) The rotational speed increases by a factor of 100.

35 mph →

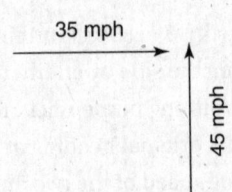

↑ 45 mph

20. Two cars approach an intersection at right angles, as pictured above. What is their relative speed to each other?

(A) 10 mph
(B) 45 mph
(C) 57 mph
(D) 80 mph

21. A container of water (pictured below) has two holes in its side allowing two streams of water to emerge as shown. Initially, the lower hole is twice as deep as the higher hole. Which statement best compares the initial velocities of these two streams of water?

(A) They emerge at the same speed.
(B) The lower stream emerges 2½ times as fast as the upper stream.
(C) The lower stream emerges 2 times as fast as the upper stream.
(D) The lower stream emerges 4 times as fast as the upper stream.

Section II: Free-Response

1. A static, nonuniform wedge-shaped mass, m (with center of mass x as indicated), with a base length of L (shown below), is supported at two points. The first support point is at the bottom left corner of the wedge. The second support point is located at a point ¾ of the length of the wedge away from that corner.

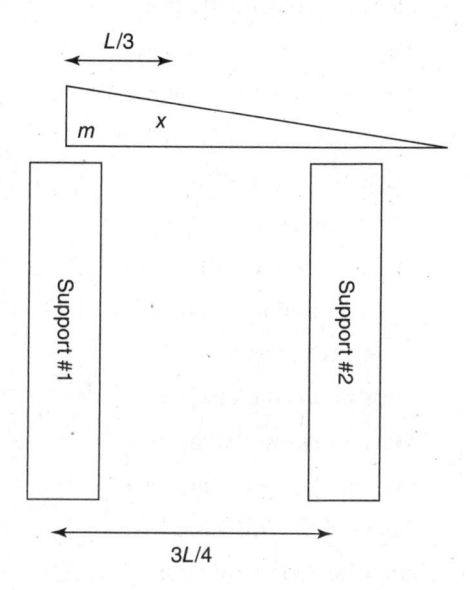

(a) Using as givens m, L, and g (the gravitational field strength), solve for the two unknown contact forces at the points of support (F_1 and F_2).
(b) Support #2 is removed. The wedge begins to fall by rotating around support #1 with an angular acceleration of 2 rad/s². If the mass of the wedge is 3.8 kg and $L = 75$ cm, determine the rotational inertia of the nonuniform wedge about the top of support #1.
(c) As the wedge continues to rotate clockwise but remains in contact with support #1:
    (i) Does the angular acceleration increase, decrease, or remain the same? Justify your response.
    (ii) Does the rotational inertia of the wedge increase, decrease, or remain the same? Justify your response.

2. An object of unknown mass is fired at an unknown angle with an initial speed of 25 m/s. Ignore frictional effects. If only 45 percent of its initial kinetic energy is still present as kinetic energy at the projectile's highest point, determine the time in flight for the projectile.

Answer Key	Topic and Chapter to Reference
1. **(C)**	Statics problem, Chapter 3
2. **(B)**	Frictional dynamics, Chapter 3
3. **(B)**	Elevator problem, Chapter 3
4. **(A)**	Newton's first law, Chapter 3
5. **(C)**	Newton's third law, Chapter 3
6. **(B)**	Friction, Chapters 2 and 3
7. **(A)**	Friction, Chapters 2 and 3
8. **(D)**	Kinematics, Chapter 2
9. **(D)**	Projectile motion, Chapter 2
10. **(C)**	Circular motion, Chapter 2
11. **(C)**	Gravity, Chapter 5
12. **(A)**	Mass vs. weight, Chapter 3
13. **(D)**	Work and power, Chapter 4
14. **(B)**	Net work = ΔKE, Chapter 4
15. **(C)**	Net work = ΔKE, Chapter 4
16. **(B)**	Impulse and momentum, Chapter 6
17. **(D)**	Conservation of momentum, Chapter 6
18. **(D)**	Torque, Chapter 7
19. **(D)**	Conservation of angular momentum, Chapter 7
20. **(C)**	Relative velocity, Chapter 2
21. **(B)**	Fluids, Chapter 9

STOP If there is still time remaining, you may review your answers.

Answers Explained

Section I: Multiple-Choice

1. **(C)** $T < mg$. Since the mass is stationary, we conclude that the net force is zero. The upward tension of the angled rope must be mg in order to cancel out gravity. The free-body diagram on the hanging mass is:

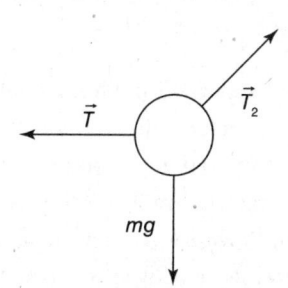

   Since the angle θ is greater than 45 degrees, the vertical component of tension T_2 must be bigger than the horizontal component. (To see this quickly, draw an extreme case, such as 85 degrees.) Therefore, the horizontal component of tension must be less than mg. The horizontal rope T must cancel this horizontal component of T_2.

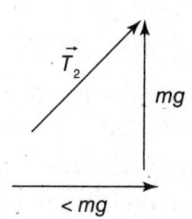

2. **(B)** $\mu > 0.8$. The normal force between the mass and the floor is 150 N. The needed sideways push to overcome static friction is **greater than** 120 N. Dividing these two forces would give us:

   $$\mu_s = 120/150 = 0.8$$

   Since a greater force is needed, the coefficient of static friction must be greater than this value.

3. **(B)** $0.5mg$. The free-body diagram consists of only two forces.

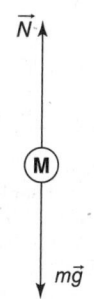

   Newton's second law gives us $+N - mg = ma$. Let $a = -0.5g$, which is half the value of g and downward:

   $$N - mg = -0.5mg$$
   $$N = 0.5mg$$

4. **(A)** The ball's inertia keeps it going. Newton's first law says that a body's inertia maintains its current motion once in motion. After the force of the throw, the only force on the ball is the vertical force of gravity. This causes the ball to accelerate in the vertical direction while the ball's sideways velocity is maintained due to its inertia.

5. **(C)** Force on hand = force on book. Newton's third law of action-reaction states that the force pair of interaction between two objects is always equal and opposite. The acceleration of either object has no role in the third law. The acceleration comes about from the net force on a single object.

6. **(B)** 0.25. To find force, we must first find acceleration:

   $$v_f^2 = v_i^2 + 2ad$$
   $$(0)^2 = (10 \text{ m/s})^2 + (2a)(20 \text{ m})$$
   $$a = -2.5 \text{ m/s}^2$$

The net force horizontally is the friction. There are no other horizontal forces:

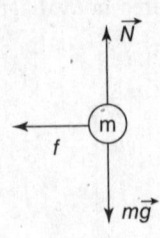

Newton's second law gives us:

$$f = ma = -m(2.5)$$

Our model for friction gives us:

$$f = -\mu N$$

This value is negative since it is to the left. Since the vertical forces cancel out, $N = mg$. Substituting for friction from above and mg for N:

$$-m(2.5) = -\mu mg$$

Note that mass cancels:

$$\mu = (2.5/g) = 0.25$$

7. **(A)** The rock will fall an additional 4.9 meters is NOT true. An object in free fall will accelerate by a constant 9.8 m/s². Since the rock is already moving downward, this leads to an increase in speed of 9.8 m/s in that 1-second interval. However, the rock will fall much farther than 4.9 m as it is already moving close to 30 m/s downward by the start of the 3rd second. It will fall an additional 35 meters in that 4th second. The rock does indeed fall 4.9 meters in the 1st second of the drop but falls an increasingly greater distance each second thereafter.

8. **(D)** a_x is constant. All the equations of motion are predicated on the assumption of constant acceleration. Any permutation of signs or values will work in any of the equations of motion. If the acceleration is not constant, one must use graphical methods or approximate the acceleration as constant over a narrow interval of time.

9. **(D)** They are all the same except for V_y. Projectile motion involves a constant vertical acceleration of $-g$ and no horizontal acceleration at all. Therefore, a_x and a_y are both constant throughout the flight. Since a_x is zero, there is no change in V_x. So V_x is also constant. V_y, however, is always changing. For a symmetric problem such as this one, the projectile lands with the same vertical speed as it had initially but in the opposite direction so that $V_{yf} = -V_{yi}$.

10. **(C)** The bucket will fly out horizontally initially. Objects executing circular motion have velocity vectors tangent to their motion while experiencing an inward acceleration. If the inward force is released, Newton's first law dictates that the object follow its velocity vector. Note that gravity will cause the object to fall after it is released, but the initial motion is horizontal.

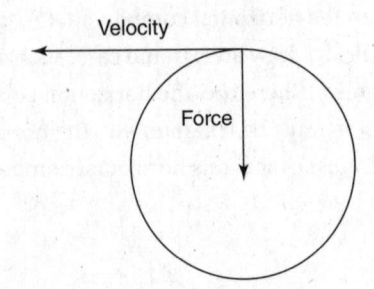

11. **(C)** The Moon's orbital period would increase by a factor of $2\sqrt{2}$. Kepler's third law of orbital bodies states that their orbital periods squared are proportional to their orbital radius cubed:

$$T_{old}^2/R_{old}^3 = T_{new}^2/R_{new}^3$$

Substituting $R_{new} = 2R_{old}$:

$$T_{old}^2/R_{old}^3 = T_{new}^2/(2R_{old})^3$$

Cancel R and solve for T_{new}:

$$T_{new} = 8^{1/2} T_{old} = 2\sqrt{2}\, T_{old}$$

Note that you can also obtain this result by examining the equation for orbital speed from Newton ($v = (Gm/r)^{1/2}$). This shows that the orbital speed of the Moon must decrease by the square root of 2. Coupling this with the knowledge that the new orbit is now twice as large, it will take $2(2^{1/2})$ or $2\sqrt{2}$ times as long to orbit once around Earth as it used to.

12. **(A)** Your mass would be the same, but your weight would change. Mass is a property of an object and measures the object's inertia regardless of location. Weight is a force due to gravity and depends both on the object's mass and on the planetary gravity the object is currently experiencing.

13. **(D)** Same work, twice the power.

$$\text{Work} = \text{force} \times \text{distance}$$

For lifting operations, the force is assumed to be equal to the weight of the object for most of the lift. (Briefly, at the beginning of the lift, the lifting force must be greater than the weight in order to get the object moving. This brief force must be different in the second case in order to get the suitcase moving faster. However, this initial force is generally ignored in these calculations as it does not happen over an appreciable distance, just as the decrease in lifting force at the end of the lift is likewise ignored.) The same force over the same distance yields the same work. If the details of the forces contributing to the work are troublesome, you may prefer to think about the energy gained by the suitcase ($W = \Delta E$). Since the suitcase has gained the same amount of energy in both cases by being lifted, the work done must likewise be the same:

$$\text{Power} = \text{work/time}$$

If the same work is done in half the time, twice the power must be supplied.

14. **(B)** $2(2)^{1/2}$ m/s. From the work-energy theorem:

$$W_{net} = \Delta KE$$

Work done by the horizontal force
$= (12 \text{ N})(5 \text{ m}) = 60 \text{ J}$
Work done by friction $= -40 \text{ J}$
$W_{net} = +60 - 40 = 20 \text{ J}$

So the kinetic energy of the mass has gone up by 20 J. Initially, it had no *KE* since it was stationary. Therefore:

$$\frac{1}{2}mv_f^2 = 20$$
$$v_f^2 = 40/5 = 8$$
$$v_f = 2(2^{1/2})$$

15. **(C)** 11.1 m/s. Mechanical energy is conserved since there are no nonconservative forces involved:

$$ME_i = ME_f$$
$$KE_i + PE_i = KE_f + PE_f$$
$$\frac{1}{2}m(2 \text{ m/s})^2 + mg(9 \text{ m}) = \frac{1}{2}mv_f^2 + mg(3 \text{ m})$$

Canceling mass in every term and solving for v_f yields:

$$v_f^2 = 2^2 + 2 \times (9g - 3g) = 4 + 12g = 124$$
$$v_f = 11.1 \text{ m/s}$$

16. **(B)** 1.75

$$\text{Impulse} = \Delta \text{momentum}$$
$$= m\vec{v}_f - m\vec{v}_i$$

Since the collision is elastic, no kinetic energy is lost. Assuming the wall remains stationary, the rubber ball must rebound with the same speed but in the opposite direction:

$$\vec{v}_f = -\vec{v}_i$$

Substituting for v_f above:

$$\text{Impulse} = m(-\vec{v}_i) - m\vec{v}_i = -2m\vec{v}_i$$
$$= -2(0.25 \text{ kg})(3.5 \text{ m/s}) = -1.75 \text{ kg} \cdot \text{m/s}$$

Here the negative sign indicates that the impulse delivered to the ball is in the opposite direction from the ball's initial velocity. Note the ball and the wall receive equal and opposite impulses (from Newton's third law) and the problem asks for magnitude (absolute value).

Since $F\Delta t$ is also an impulse, units of N · s must be the same as kg · m/s and is quickly shown:

$$\text{N} \cdot \text{s} = (\text{kg} \cdot \text{m/s}^2)\text{s} = \text{kg} \cdot \text{m/s}$$

17. **(D)** 1.1 m/s. Total momentum must be conserved through the collision. So we need to find only the magnitude of the initial momentum of the system. Each player is originally running perpendicularly to the other. Thus, their individual momentums are different components. Imagine one as an x-component and the other as a y-component. Therefore, the resulting magnitude is found using the Pythagorean theorem. Dividing by total mass yields the speed of the combined players after impact:

$$p_{1x} = 85(1.5) = 127.5 \text{ kg} \cdot \text{m/s}$$
$$p_{2y} = 75(1.75) = 131.25 \text{ kg} \cdot \text{m/s}$$
$$\text{Magnitude} = 183 \text{ kg} \cdot \text{m/s}$$
$$\text{Net mass} = 160 \text{ kg}$$
$$\text{Speed} = \text{magnitude/mass} = 1.14 \text{ m/s}$$

18. **(D)** $\tau_B > \tau_A > \tau_C$

    Torque = (force)(distance from axis of rotation) $\sin\theta$

    Force C is applied along the lever arm. So force C has an angle of 180°, resulting in no torque.

    Force A has a much shorter lever arm than force B. Therefore, the torque supplied by A is much less than the torque supplied by B.

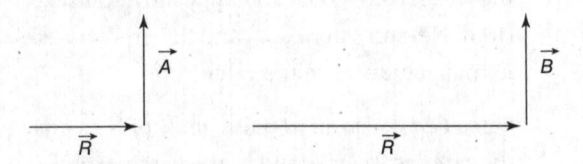

19. **(D)** The rotational speed increases by a factor of 100. The rotational inertia (I) for a sphere is proportional to R^2. Because the shape of the cloud is not changing, we can determine that the rotational inertia has decreased by a factor of 10^2. Since angular momentum is conserved:

    $$I_1\omega_1 = I_2\omega_2$$

    The new angular velocity (ω_2) must be 100 times bigger:

    $$I_1\omega_1 = I_2\omega_2 = (I_1/100)(100\,\omega_1)$$

20. **(C)** 57 mph. Relative velocity is the difference in the velocity vectors. When the vectors are subtracted, the resultant is the hypotenuse of a right triangle:

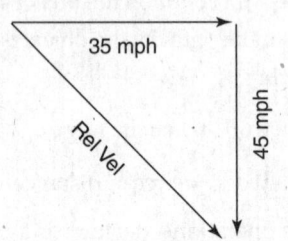

    Relative velocity = $(35^2 + 45^2)^{½}$

21. **(B)** According to Bernoulli's equation, the square of the velocity of fluid relates to the depth of the fluid, assuming the pressure is the same. (This occurs since the streams of water are released into equal atmospheric pressure and is also known as Torricelli's theorem.) Comparing the surface of the water to one of the streams:

    $$P_1 + \rho g y_1 + \tfrac{1}{2}\rho v_1^2 = P_2 + \rho g y_2 + \tfrac{1}{2}\rho v_2^2$$

    With P_1 and P_2 being equal and the surface being stationary ($v_1 = 0$), we are left with:

    $$\rho g y_1 = \rho g y_2 + \tfrac{1}{2}\rho v_2^2$$

    Eliminating and rearranging gives:

    $$v_2^2 = 2g(y_1 - y_2)$$

    Since the lower hole is twice as deep, the quantity in parentheses will be twice as great for the lower hole as compared with the first. To compare velocities, take the square root of both sides.

Section II: Free-Response

1. This is a statics problem (see Chapters 3 and 7 for more information).

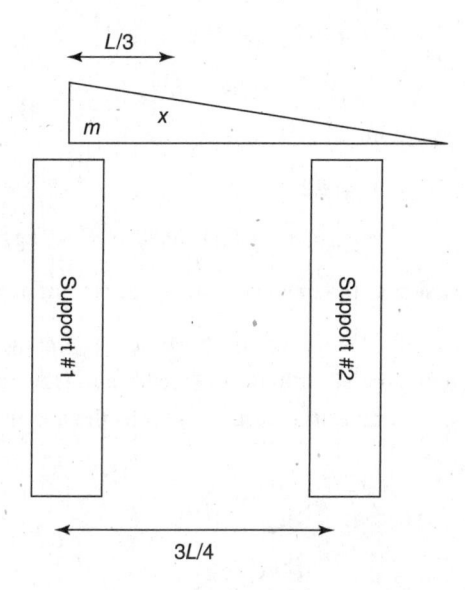

    (a) Both net force and net torque must add up to zero. Make a free-body diagram:

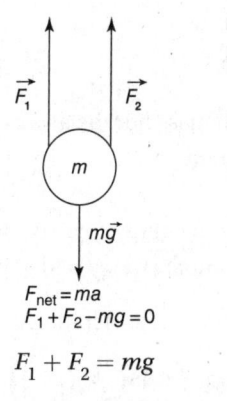

    $F_{net} = ma$
    $F_1 + F_2 - mg = 0$

    $$F_1 + F_2 = mg$$

    We now have two unknowns but only one equation! For torque, we are free to choose any axis of rotation we want since the wedge is not actually moving. We should place the axis of rotation at either contact point in order to eliminate one variable. If we place our axis of rotation at the F_1 contact point and use minus for clockwise rotations and plus for counterclockwise rotations:

    $$\text{Net Torque} = I\alpha$$
    $$F_1(0) - mg(L/3) + F_2(3L/4) = 0$$
    $$F_2(3L/4) = mg(L/3)$$
    $$F_2 = 4mg/9$$

    Combine this equation with the one found above.

    $$F_1 = mg - 4mg/9 = 5mg/9$$

(b) $$\text{Torque} = I\alpha$$

Since F_2 is gone, the torque is supplied only by the mass of the object:

$$\text{Torque} = FR \sin\theta$$
$$\text{Torque} = mg(L/3)(1)$$

Substituting into the above equation:

$$mgL/3 = I\alpha = I(2)$$
$$I = mgL/6 = (3.8)(9.8)(0.75)/6 = 4.7 \text{ kg}\cdot\text{m}^2$$

Note that $I = mr^2$ cannot be used, as the object has mass spread out over many different r values.

(c) (i) As the object rotates, neither the lever arm nor the force (mg) changes. However, the angle between the lever arm and force goes from 90° initially toward 0° at the lowest point of the downward swing. Since torque is proportional to $\sin\theta$, the torque (and thus the angular acceleration) will decrease.

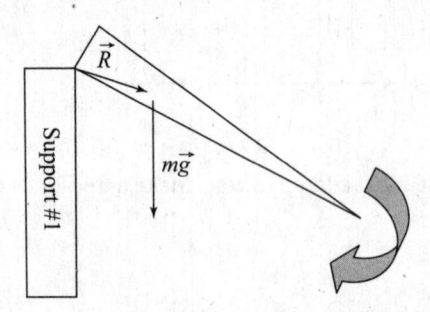

(ii) Since neither the distribution of mass nor the axis of rotation changes during the swing, the object's rotational inertia remains constant.

2. This is a projectile motion (Chapter 2) and energy conservation (Chapter 4) problem. Energy conservation tells us that the other 55 percent of the initial energy is in gravitational potential energy.

$$\text{GPE }(U_s) \text{ at top} = 55\% \text{ of initial } KE = 0.55\left(\tfrac{1}{2}m(25^2)\right)$$
$$mgh = 0.55\left(\tfrac{1}{2}m(25^2)\right)$$

Solve for height and note that mass cancels:

$$h = 17.5 \text{m}$$

Determine the time to fall from that height. Remember there is no vertical velocity at the highest point!

$$D_y = \tfrac{1}{2}at^2$$
$$17.5 = 4.9t^2$$
$$t = 1.89 \text{ s}$$

Double this value to obtain time in flight: 3.8 seconds.

Review and Practice

1

Vectors

Learning Objectives

In this chapter, you will learn about:
→ Coordinate systems and frames of reference
→ Vectors
→ Addition of vectors
→ Subtraction of vectors
→ Addition methods using the components of vectors

Coordinate Systems and Frames of Reference

We begin our review of physics with the idea that all observations and measurements are made relative to a suitably chosen frame of reference. In other words, when observations are made that will be the basis of future predictions, we must be careful to note from what point of view those observations are being made. For example, if you are standing on the street and see a car driving by, you observe the car and all of its occupants moving relative to you. However, to the driver and other occupants of the car, the situation is different: they may not appear to be in motion relative to themselves; rather, you appear to be moving backward relative to them.

If you were to get into your car, drive out to meet the other car, and travel at the same speed in the same direction right next to the first car, there would be no relative motion between the two cars. These different points of view are known as **frames of reference**, and they are very important aspects of physics.

A coordinate system within a frame of reference is defined to be a set of reference lines that intersect at an arbitrarily chosen fixed point called the **origin**. In the Cartesian coordinate system, the reference lines are three mutually perpendicular lines designated x, y, and z (see Figure 1.1). The coordinate system must provide a set of rules for locating objects within that frame of reference. In the three-dimensional Cartesian system, if we define a plane containing the x- and y-coordinates (let's say a horizontal plane), then the z-axis specifies direction up or down.

It is often useful to compare observations made in two different frames of reference. In the example above, the motion of the occupants of the car was reduced to zero if we transformed our coordinate system to the car moving at constant velocity. In this case, we say that the car is an **inertial frame of reference**. In such a frame, it is impossible to observe whether or not the reference frame is in motion if the observers are moving with it.

> **REMEMBER**
>
> A frame of reference represents an observer's viewpoint and requires a coordinate system to set the origin. A frame of reference moving with a constant velocity is called an inertial frame of reference.

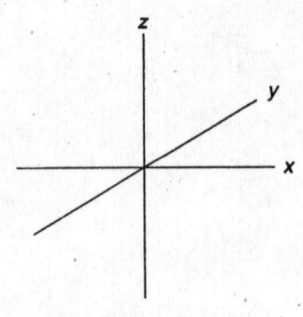

Figure 1.1

Vectors

Another way of locating the position of an object in the Cartesian system is with a directed line segment, or **vector**. If you draw an arrow, starting from the origin to a point in space (see Figure 1.2), you have defined a position vector $\vec{R}$, whose magnitude is given by $|\vec{R}|$ and is equal to the linear distance between the origin and the point (x, y). The direction of the vector is given by the angle, identified by the Greek letter θ, that the arrow makes with the positive x-axis. Any quantity that has both magnitude and direction is called a **vector quantity**. Any quantity that has only magnitude is a **scalar quantity**. Examples of vector quantities are force, velocity, weight, and displacement. Examples of scalar quantities are mass, distance, speed, and energy.

> **TIP**
> Vectors have both magnitude (size) and direction. Scalars have only magnitude.

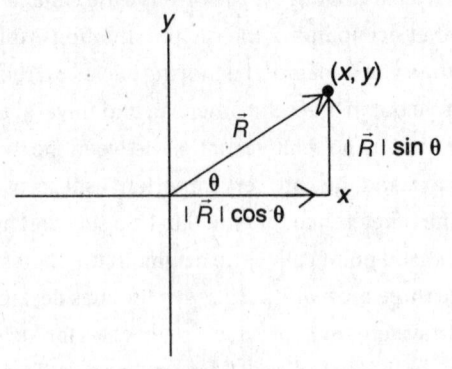

Figure 1.2

If we designate the magnitude of the vector $\vec{R}$ as r, and the direction angle is given by θ, then we have an alternative coordinate system called the **polar coordinate system**. In two dimensions, to locate the point in the Cartesian system (x, y) involves going x units horizontally and y units vertically. Additionally, we can show that (see Figure 1.2)

$$x = r\cos\theta \quad \text{and} \quad y = r\sin\theta$$

Having established this polar form for vectors, we can easily show that r and θ can be expressed in terms of x and y as follows:

$$r^2 = x^2 + y^2$$
$$\tan\theta = \frac{y}{x}$$

The magnitude of $\vec{R}$ is then given by

$$|\vec{R}| = r = \sqrt{x^2 + y^2}$$

and the direction angle θ is given by

$$\theta = \tan^{-1}\frac{y}{x}$$

Here is an example of a vector described in two equivalent ways:

1. **MAGNITUDE AND DIRECTION:** useful for drawing, visualizing, and graphical analysis
2. **COMPONENT FORM:** useful for calculations and mathematical operations

You must be able to use the math in this section to go back and forth between these two representations fluently.

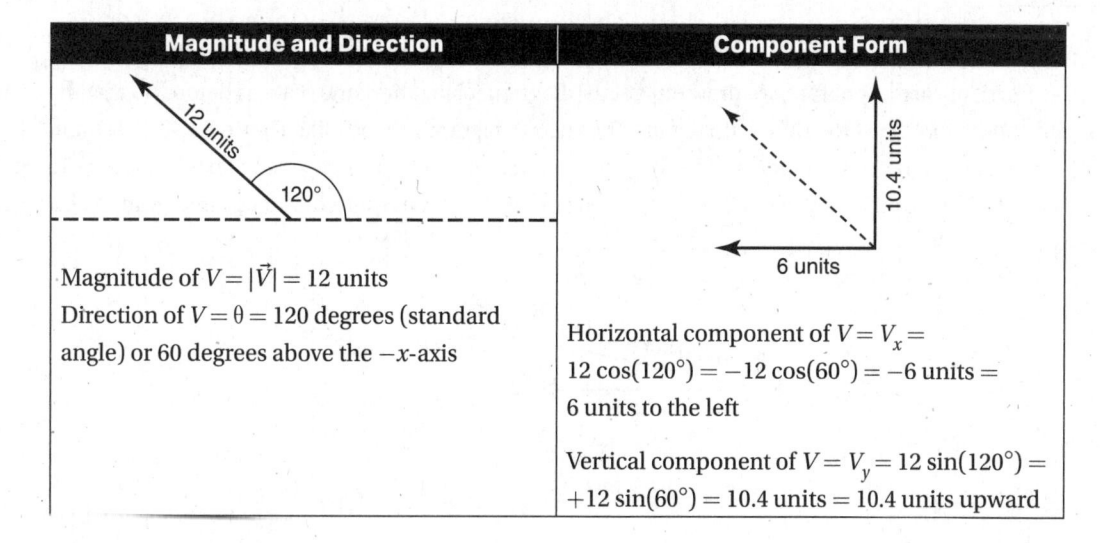

Magnitude and Direction	Component Form
12 units at 120°	6 units horizontal, 10.4 units vertical
Magnitude of $V = |\vec{V}| = 12$ units	Horizontal component of $V = V_x = 12\cos(120°) = -12\cos(60°) = -6$ units = 6 units to the left
Direction of $V = \theta = 120$ degrees (standard angle) or 60 degrees above the $-x$-axis	Vertical component of $V = V_y = 12\sin(120°) = +12\sin(60°) = 10.4$ units = 10.4 units upward

Addition of Vectors

Geometric Considerations

The ability to combine vectors is a very important tool of physics. From a geometric standpoint, the "addition" of two vectors is not the same as the addition of two numbers. When we state that, given two vectors $\vec{A}$ and $\vec{B}$, we wish to form the third vector $\vec{C}$ such that $\vec{A} + \vec{B} = \vec{C}$, we must be careful to preserve the directions of the vectors relative to our chosen frame of reference.

One way to do this "addition" is by the construction of what is called a **vector diagram**. Vectors are identified geometrically by a directed arrow that has a "head" and a "tail." We will consider a series of examples.

REMEMBER

Vectors add constructively from head to tail. The vector sum is called the resultant.

Example 1

First, suppose a girl is walking from her house a distance of five blocks east and then an additional two blocks east. How can we represent these displacements vectorially? First, we must choose a suitable scale to represent the magnitude lengths of the vectors. In this case, let us just call the scale "one vector unit" or just "one unit," to be equal to one block of distance. Let us also agree that "east" is to the right (and hence "north" is directed up). A vector diagram for this set of displacements is given in Figure 1.3. Note that the resultant has both magnitude and direction as it, too, is a vector.

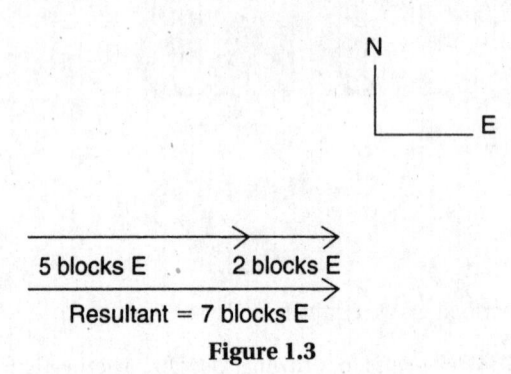

Figure 1.3

Example 2

Now, for our second example, suppose the girl walks five blocks east and then two blocks northeast (that is, 45 degrees north of east). As before, we draw our vector diagram, using the same scale as before, so that the two vectors are connected "head to tail" or "tip to tail." The vector diagram will look like Figure 1.4. The resultant is drawn from the tail of the first vector to the head of the second, forming a triangle (some texts use the "parallelogram" method of construction, which is equivalent). The direction of the resultant is measured from the horizontal axis as usual.

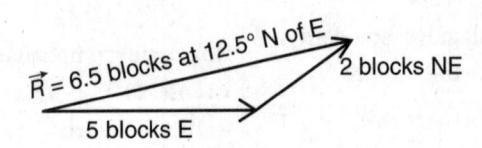

Figure 1.4

Example 3

For our third example, suppose the girl walks five blocks east and then two blocks north. Figure 1.5 shows the vector diagram for this set of displacements.

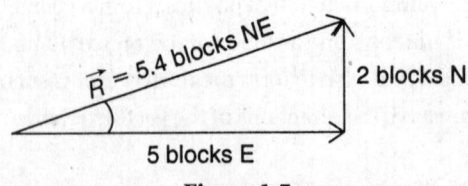

Figure 1.5

> **Example 4**

Finally, for our fourth example, suppose the girl walks five blocks east and then two blocks west. Her final displacement will be three blocks east of the starting point as shown in Figure 1.6.

Figure 1.6

In our four examples (Figures 1.3, 1.4, 1.5, and 1.6), the two vectors of magnitudes 5 and 2 could add to any value between 7 and 3, depending on their relative orientation.

From these examples, we conclude that, as the angle between two vectors increases, the magnitude of their resultant decreases. Also, the magnitude of the resultant between two vectors is a maximum when the vectors are in the same direction (at a relative angle of 0 degrees) and a minimum when they are in the opposite direction (at a relative angle of 180 degrees). It is important to remember that the vectors must be constructed head to tail and a suitable scale chosen for the system.

Algebraic Considerations

In the above examples, the resultant between two vectors was constructed using a vector diagram. The magnitude of the resultant was the measured length of the vector drawn from the tail of the first vector to the head of the second. If we sketch such a situation, we form a triangle whose sides are related by the **law of cosines** and whose angles are related by the **law of sines**.

Consider the vector triangle in Figure 1.7 with arbitrary sides a, b, and c and corresponding angles A, B, and C.

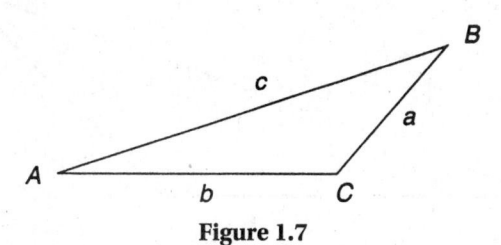

Figure 1.7

The law of cosines states that

$$c^2 = a^2 + b^2 - 2ab \cos C$$

The law of sines states that

$$\frac{a}{\sin A} = \frac{b}{\sin B} = \frac{c}{\sin C}$$

If we use the information given in our second example above, we see that in Figure 1.8 the vectors have the following magnitudes and directions:

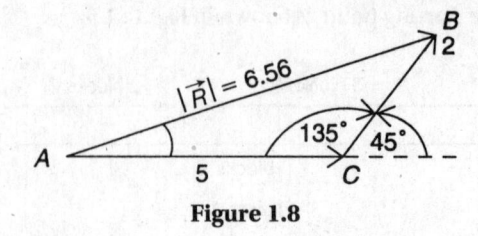

Figure 1.8

Using the law of cosines, we obtain the magnitude of the resultant in "blocks":

$$c = \sqrt{(2)^2 + (5)^2 - 2(2)(5)\cos 135} = 6.56$$

Using the law of sines, we find that the direction of the resultant is angle A:

$$\frac{6.56}{\sin 135} = \frac{2}{\sin A}$$

Angle A turns out to be equal to 12.45 degrees north of east.

Addition of Multiple Vectors

As long as the vectors are constructed head to tail, multiple vectors can be added in any order. If the resultant is zero, the diagram constructed will be a closed geometric figure. This can be shown in the case of three vectors forming a closed triangle, as in Figure 1.9.

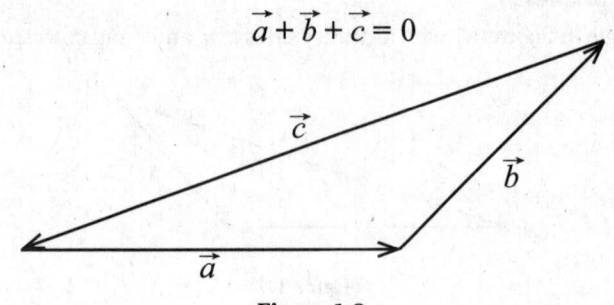

Figure 1.9

Thus, if the girl walks five blocks east, two blocks north, and five blocks west, the vector diagram will look like Figure 1.10, with the resultant displacement being equal to two blocks north.

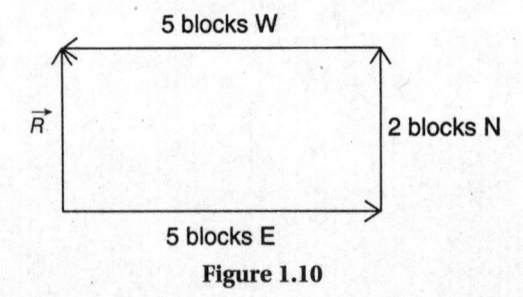

Figure 1.10

Subtraction of Vectors

Subtraction of vectors is best understood by rethinking the subtraction as the addition of a negative vector.

$$\vec{R}_2 - \vec{R}_1 = \vec{R}_2 + (-\vec{R}_1)$$

A negative vector is simply one of the same magnitude but directed opposite (i.e., a 180-degree rotation). Note that the result of this operation is the same here as in the example that follows since the resultant has the same magnitude and direction. See Figures 1.11 and 1.12. This difference between the two vectors is sometimes called $\Delta\vec{R}$, so that $\Delta\vec{R} = \vec{R}_2 - \vec{R}_1$.

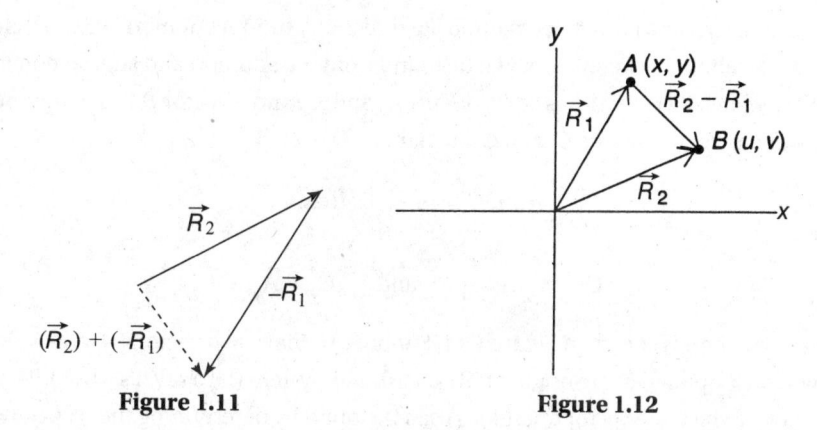

Figure 1.11

Figure 1.12

From our understanding of vector diagrams, we can see that $\vec{R}_1$ and $\Delta\vec{R}$ are connected head to tail, and thus that $\vec{R}_1 + \Delta\vec{R} = \vec{R}_2$!

Addition Methods Using the Components of Vectors

Any single vector in space can be **resolved** into two perpendicular components in a suitably chosen coordinate system. The methods of vector resolution were discussed at the beginning of this section, but we can review them here. Given a vector (sketched in Figure 1.13) representing a displacement of 100 meters northeast (that is, making a 45-degree angle with the positive *x*-axis), we can observe that it is composed of an *x*- and a *y*- (a horizontal and a vertical) component that can be geometrically or algebraically determined.

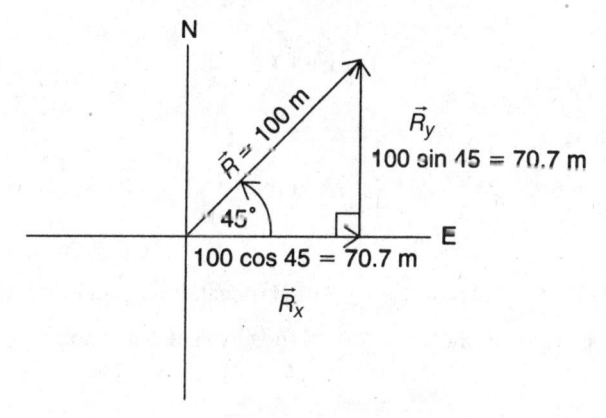

Figure 1.13

Geometrically, if we were to draw the 100-meter vector to scale at the correct angle, then projecting a perpendicular line down from the head of the given vector to the x-axis would construct the two perpendicular components. Algebraically, we see that if we have a given vector $\vec{R}$, then, in the x-direction, we have

$$R_x = |\vec{R}| \cos \theta \quad \text{and} \quad R_y = |\vec{R}| \sin \theta$$

where the magnitude of $\vec{R}$ is R. The magnitude R may also be written as $|\vec{R}|$.

This method of vector resolution can be useful when adding two or more vectors. Since all vectors in the same direction add up numerically, and all vectors in opposite directions subtract numerically, therefore, if we are given two vectors, the resultant between them will be found from the addition and subtraction of the respective components. In other words, if vector $\vec{A}$ has components A_x and A_y, and if vector $\vec{B}$ has components B_x and B_y, the resultant vector $\vec{C}$ will have components C_x and C_y such that

$$\text{if } \vec{C} = \vec{A} + \vec{B}, \text{ then}$$

$$C_x = A_x + B_x \quad \text{and} \quad C_y = A_y + B_y$$

An example of this method is shown in Figure 1.14. Suppose we have two vectors, $\vec{A}$ and $\vec{B}$, that represent tensions in two ropes applied at the origin of the coordinate system (let's say it's a box). The resultant force between these tensions (which are vector quantities) can be found by determining the respective x- and y-components of the given vectors.

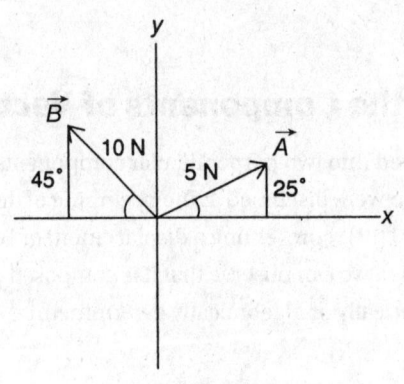

Figure 1.14

From the diagram, we see that

$$A_x = 5 \cos 25 = 4.53 \text{ N} \quad \text{and} \quad A_y = 5 \sin 25 = 2.11 \text{ N}$$

For the second vector, we have

$$B_x = -10 \cos 45 = 10 \cos 135 = -7.07 \text{ N (negative since it is pointing left)}$$
$$B_y = 10 \sin 45 = 10 \sin 135 = 7.07 \text{ N (positive since it is pointing up)}$$

Therefore,

$$C_x = 4.53 \text{ N} - 7.07 \text{ N} = -2.54 \text{ N} \quad \text{and} \quad C_y = 2.11 \text{ N} + 7.07 \text{ N} = 9.18 \text{ N}$$

The magnitude of the resultant vector, $\vec{C}$, is given by

$$|\vec{C}| = \sqrt{(-2.54)^2 + (9.18)^2} = 9.52 \text{ N}$$

$\tan^{-1}(9.18/-2.54) \doteq -74.5$

Although calculators always return values in the first or fourth quadrant, you must correct this when the vector is known to be in the third or second quadrant. Accomplish this by adding 180 degrees. Judging by the signs of its component $(-,+)$, this vector C is in the second quadrant.

The angle that vector $\vec{C}$ makes with the positive x-axis is about $105°$.

Tip-to-tail method:

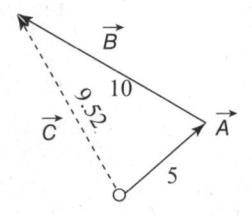

SUMMARY

- Vectors are quantities that have both magnitude (size) and direction.
- Vectors can be resolved into components that represent the magnitude and direction of a vector along a particular axis.
- Scalars are quantities without direction.
- Force, displacement, and velocity are examples of vectors.
- Distance, speed, and mass are examples of scalars.
- Vectors can be "added" geometrically by the tip-to-tail method.
- The resultant of two vectors is equal to the vector obtained by "adding" them.
- If two vectors are in the same direction, their resultant is equal to the sum of their magnitudes.
- If two vectors are in the opposite direction, their resultant is equal to the difference of their magnitudes.

Problem-Solving Strategies for Vectors

When solving a vector problem, be sure to:

1. Understand the frame of reference for the situation.
2. Select an appropriate coordinate system for the situation. Cartesian systems do not necessarily have to be vertical and horizontal (a mass sliding down an inclined plane is an example of a rotated system with the "x-axis" parallel to the incline).
3. Pick an appropriate scale for starting your vector diagram. For example, in a problem with displacement, a scale of 1 cm = 1 m might be appropriate.
4. Recognize the given orientation of the vectors as they correspond to geographical directions (N, E, S, W). Be sure to maintain the same directions as you draw your vector triangle.
5. Connect vectors head to tail, and connect the resultant from the tail of the first vector to the head of the last vector.
6. Identify and determine the components of the vectors. In most cases, vector problems can be simplified using components.

Practice Exercises

Multiple-Choice

1. A vector is given by its components, $A_x = 2.5$ and $A_y = 7.5$. What angle does vector $\vec{A}$ make with the positive x-axis?

   (A) 72°
   (B) 18°
   (C) 25°
   (D) 50°

2. Which pair of vectors could produce a resultant of 35?

   (A) 15 and 15
   (B) 20 and 20
   (C) 30 and 70
   (D) 20 and 60

3. A vector has a magnitude of 17 units and makes an angle of 20° with the positive x-axis. The magnitude of the horizontal component of this vector is

   (A) 16 units
   (B) 4.1 units
   (C) 5.8 units
   (D) 50 units

4. As the angle between a given vector and the horizontal axis increases from 0° to 90°, the magnitude of the vertical component of this vector

   (A) decreases
   (B) increases and then decreases
   (C) decreases and then increases
   (D) increases

5. Vector $\vec{A}$ has a magnitude of 10 units and makes an angle of 30° with the horizontal x-axis. Vector $\vec{B}$ has a magnitude of 25 units and makes an angle of 50° with the negative x-axis. What is the magnitude of the resultant of these two vectors?

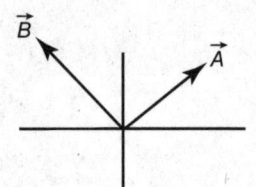

   (A) 20
   (B) 35
   (C) 15
   (D) 25

6. Two concurrent vectors have magnitudes of 3 units and 8 units. The difference between these vectors is 8 units. The angle between these two vectors is

   (A) 34°
   (B) 56°
   (C) 79°
   (D) 113°

7. Which of the following sets of displacements have equal resultants when performed in the order given?

   I. 6 m east, 9 m north, 12 m west
   II. 6 m north, 9 m west, 12 m east
   III. 6 m east, 12 m west, 9 m north
   IV. 9 m north, 6 m east, 12 m west

   (A) I and IV
   (B) I and II
   (C) I, III, and IV
   (D) I, II, and IV

8. Which vector represents the direction of the two concurrent vectors shown below?

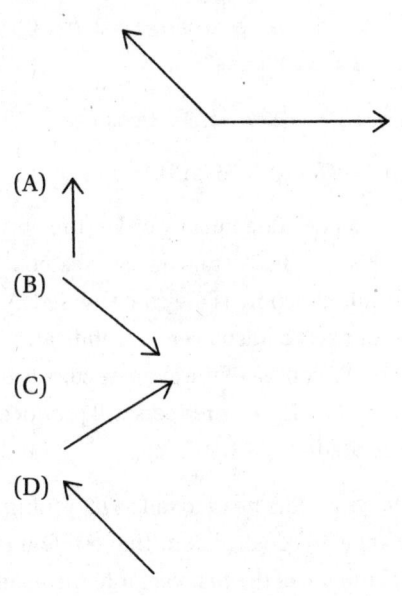

(A)
(B)
(C)
(D)

9. Three forces act concurrently on a point P as shown below. Which vector represents the direction of the resultant force on point P?

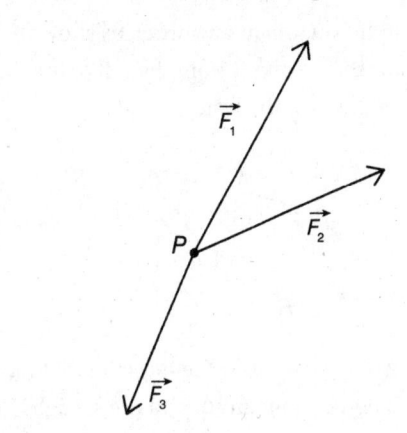

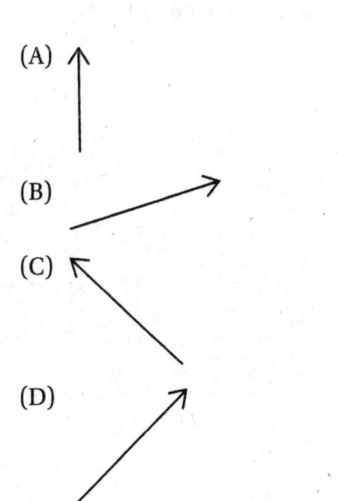

(A)
(B)
(C)
(D)

10. On a baseball field, first base is about 30 m away from home plate. A batter gets a hit and runs toward first base. She runs 3 m past the base and then runs back to stand on it. The magnitude of her final displacement from home plate is

(A) 27 m
(B) 30 m
(C) 33 m
(D) 36 m

Additional Practice

1. Find the magnitude and direction of the two concurrent forces shown below using the algebraic method of components.

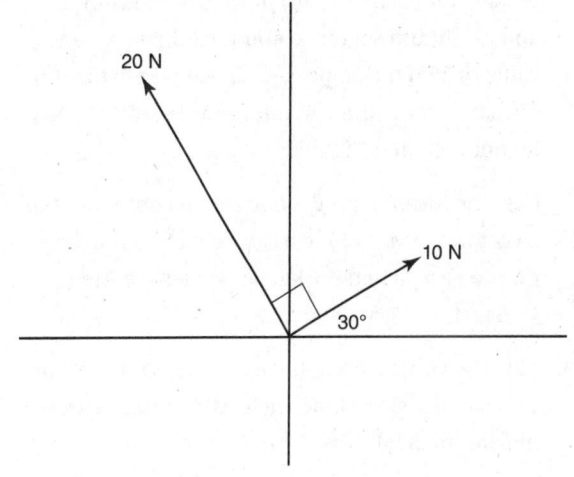

Note: Figure is not drawn to scale.

2. Two vectors, $\vec{A}$ and $\vec{B}$, are concurrent and attached at the tails with an angle θ between them. Given that each vector has components A_x, A_y and B_x, B_y, respectively, use the law of cosines to show that

$$\cos\theta = \frac{A_x B_x + A_y B_y}{|\vec{A}||\vec{B}|}$$

3. Give a geometric explanation for the following statement: "Three vectors that add up to zero must be coplanar (that is, they must all lie in the same plane)."

4. (a) Is it necessary to specify a coordinate system when adding two vectors?
   (b) Is it necessary to specify a coordinate system when forming the components of a vector?

5. Can a vector have zero magnitude if one of its components is nonzero?

Answers Explained

Multiple-Choice Problems

1. **(A)** The angle for the vector to the positive x-axis is given by

    $$\tan \theta = \frac{A_y}{A_x} = \frac{7.5}{2.5}$$

    Thus, $\tan \theta = 3$ and $\theta = 71.5°$ or, rounded off, $72°$.

2. **(B)** Two vectors have a maximum resultant, whose magnitude is equal to the numerical sum of the vector magnitudes, when the angle between vectors is $0°$. The resultant is a minimum at $180°$, and the magnitude is equal to the numerical difference between the magnitudes. When this logic is used, only pair (20, 20) produces a maximum and minimum set that could include the given value of 35 if the angle were specified. All the others have maximum and minimum resultants that do not include 35 in their range.

3. **(A)** The formula for the horizontal component of a vector $\vec{A}$ is $A_x = |\vec{A}| \cos \theta$, where $|\vec{A}|$ is the magnitude of $\vec{A}$. When the known values are used, values $A_x = 17 \cos 20 = 15.9$ units.

4. **(D)** The vertical component of a vector is proportional to the sine of the angle, which increases to a maximum at $90°$.

5. **(D)** Form the components of each vector and then add these components, remembering the sign conventions; right and up are positive values. Thus, we have $A_x = 10 \cos 30 = 8.66$ and $A_y = 10 \sin 30 = 5$, $B_x = -25 \cos 50 = -16.06$ and $B_y = 25 \sin 50 = 19.15$.

    The components of the resultant are $C_x = A_x + B_x = 8.66 - 16.06$, so $C_x = -7.4$ units and $C_y = A_y + B_y = 5 + 19.15 = 24.15$. The magnitude of this resultant vector is given by

    $$|\vec{C}| = \sqrt{C_x^2 + C_y^2} = \sqrt{(7.4)^2 + (24.15)^2}$$

    which rounds to $|\vec{C}| = 25$ units.

6. **(C)** Use the law of cosines where $a = 3$, $b = 8$, and $c = 8$, and solve for $\cos \theta$:

    $$(8)^2 = (3)^2 + (8)^2 - 2(3)(8)\cos \theta$$

    Solving gives $\cos \theta = 0.1875$ and $\theta = 79°$.

7. **(C)** Vectors can be added in any order. The only requirement is that the vectors have the same magnitude and direction as they are shuffled. A look at the four sets of displacements indicates that I, III, and IV consist of the same vectors listed in different orders. These three sets will produce the same resultant.

8. **(C)** Sketch the vectors head to tail as if forming a vector triangle in construction. The resultant is drawn from the tail of the first vector to the head of the second. Choose the horizontal vector as the first one; then choice (C) is the general direction of the resultant.

9. **(B)** Sketch the vectors in any order, as shown below. Draw the resultant from the tail of the first vector to the head of the last.

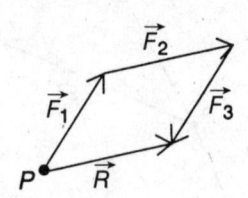

10. **(B)** The displacement is in the direction of first base and is equal in magnitude to the straight-line distance between home plate and first base (30 m).

Additional Practice

1. Let $A = 10$ N force and $B = 20$ N force.

   We declare left to be negative and up to be positive.

   From the diagram, the components are:

   $$A_x = 10 \text{ N} \cos 30° = 8.66 \text{ N} \qquad B_x = -20 \text{ N} \cos 60° = -10 \text{ N}$$
   $$A_y = 10 \text{ N} \sin 30° = 5 \text{ N} \qquad B_y = 20 \text{ N} \sin 60° = 17.32 \text{ N}$$

   The components of the resultant are now given by the following equations. Watch your signs!

   $$R_x = A_x + B_x = 8.66 \text{ N} + (-10\text{N}) = -1.34 \text{ N}$$
   $$R_y = A_y + B_y = 5 \text{ N} + 17.32 \text{ N} = 22.32 \text{ N}$$

   Use the Pythagorean theorem to find the magnitude of the resultant:

   $$|\vec{R}|^2 = (-1.34 \text{ N})^2 + (22.32 \text{ N})^2$$
   $$|\vec{R}| = 22.36 \text{ N}$$

   To find the direction, notice that the resultant will lie in the second quadrant since R_x is negative and R_y is positive:

   $$\tan\theta = \frac{22.32 \text{ N}}{-1.34 \text{ N}} = -16.66$$
   $$\theta = 86.6° \text{ (relative to the } x\text{-axis)}$$

2. The situation is shown below. The "third" side of the vector triangle is $\vec{A} - \vec{B}$.

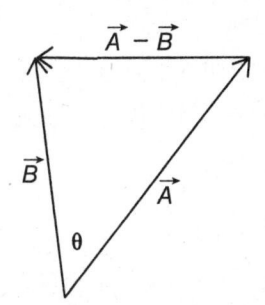

   From the rules for vector subtraction, the components of $\vec{A} - \vec{B}$ are $(A_x - B_x)$ and $(A_y - B_y)$. The law of cosines states that, for the given triangle,

   $$|\vec{A} - \vec{B}|^2 = |\vec{A}|^2 + |\vec{B}|^2 - 2|\vec{A}||\vec{B}|\cos\theta$$

   Using our definitions of magnitudes and components, we have

   $$(A_x - B_x)2 = (A_y - B_y)2 = A_x^2 + A_y^2 + B_x^2 + B_y^2 - 2|\vec{A}||\vec{B}|\cos\theta$$

   Expanding the left side and canceling like terms on both sides, we are left with

   $$-2A_xB_x - 2A_yB_y = -2|\vec{A}||\vec{B}|\cos\theta$$

   Again, canceling like terms on both sides leaves us with the final expression as we solve for $\cos\theta$:

   $$\cos\theta = \frac{A_xB_x + A_yB_y}{|\vec{A}||\vec{B}|}$$

3. Any two vectors can be added geometrically using the parallelogram method of construction. A parallelogram is a plane figure. The resultant of two vectors is represented by the diagonal of this parallelogram. In order for three vectors to add up to zero, the third remaining vector must be equal in magnitude, but opposite in direction, to the resultant of the remaining two vectors. Hence, all three must lie in the same plane to achieve this result.

4. (a) Using the logic of problem 3, any two vectors can be added by simply constructing a parallelogram from them. They can be in any orientation and in any plane. Hence, no coordinate system is necessary.
   (b) The components of a vector, by definition, lie along the axes of a chosen coordinate system (such as the x- or y-axis). Hence, to form these components, a coordinate system must be specified.

5. The magnitude of a vector is a number obtained by taking the square root of the sum of the squares of the magnitudes of its components. If one of the components is nonzero, the magnitude of the whole vector must, by definition, be nonzero as well.

2
Kinematics

Learning Objectives
In this chapter, you will learn about:
→ Average and instantaneous motion
→ Acceleration
→ Accelerated motion due to gravity
→ Graphical analysis of motion
→ Relative motion
→ Horizontally launched projectiles
→ Projectiles launched at an angle
→ Uniform circular motion

Motion involves the change in position of an object over time. When we observe an object moving, it is always with respect to a frame of reference. Since descriptions of motion have both magnitude *and* direction, there is a vector nature to motion that must be taken into account when we want to analyze how something moves.

In this chapter, we shall confine our discussions to motion in one dimension only. In physics, the study of motion is called **kinematics**. Kinematics is a completely descriptive study of *how* something moves.

Average and Instantaneous Motion

Figure 2.1

TIP
Distance is a scalar quantity, while displacement is a vector quantity.

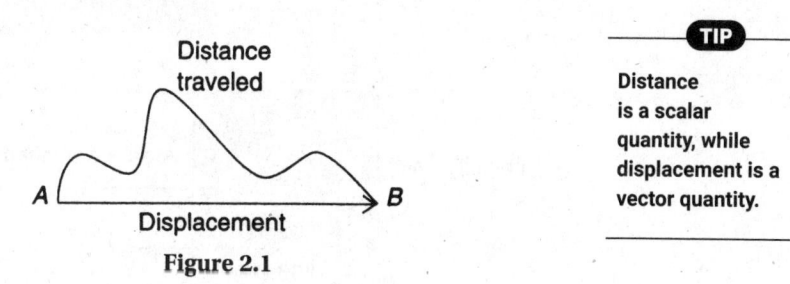

If, in Figure 2.1, we consider the actual distance traveled by the object along some arbitrary path, we are dealing with a scalar quantity. The **displacement vector** $\overline{AB}$, however, is directed along the line connecting points A and B (whether or not this is the actual route taken). Thus, when a baseball player hits a home run and runs around the bases, he or she may have traveled a distance of 360 feet (the bases are 90 feet apart), but the player's final displacement is zero (having started and ended up at the same place)!

If we are given the displacement vector of an object for a period of time Δt, we define the **average velocity**, $\vec{v}_{avg}$, to be equal to

$$\vec{v}_{avg} = \frac{\Delta \vec{x}}{\Delta t}$$

This is a vector quantity since directions are specified. Numerically, we think of the average speed as being the ratio of the total distance traveled to the total elapsed time. The units of velocity are meters per second (m/s).

If we are interested in the velocity at any instant in time, we can define the **instantaneous velocity** to be the velocity, v, as determined at any precise instant in time. In a car, the speedometer registers instantaneous speed, which can become velocity if we take into account the direction of motion. If the velocity is constant, the average and instantaneous velocities are equal, and we can write simply

$$\vec{x} = \vec{v}\Delta t$$

Note that if we were to graph time on the x-axis and displacement on the y-axis, the slope would be velocity!

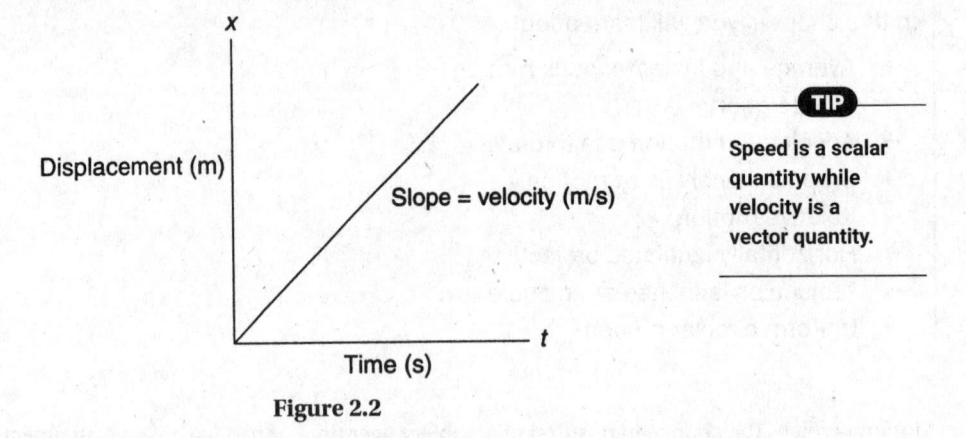

Figure 2.2

> **TIP**
> Speed is a scalar quantity while velocity is a vector quantity.

If an object is moving with a constant velocity such that its position is taken to be zero when it is first observed, a graph of the expression $\vec{x} = \vec{v}t$ would represent a direct relationship between position and time (see Figure 2.2).

Since the line is straight, the constant slope (in which $\vec{v} = \Delta\vec{x}/\Delta t$) indicates that the velocity is constant throughout the time interval. If we were to plot velocity versus time for this motion, the graph might look like Figure 2.3.

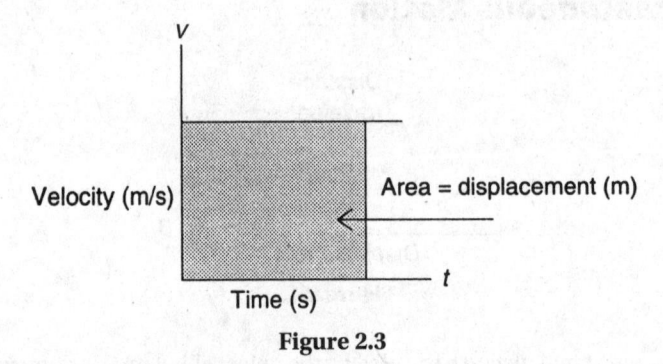

Figure 2.3

Notice that, for any time interval t, the area under the graph equals the displacement during that interval.

Acceleration

Velocity will change if either speed or direction changes. In any case, if the velocity is changing, we say that the object is **accelerating**. If the velocity is changing uniformly, the object has uniform acceleration. In this case, a graph of velocity versus time would look like Figure 2.4. Note that in Figure 2.3, the slope of the graph is zero. This makes sense since if the velocity is constant, acceleration is zero.

CHAPTER 2: KINEMATICS 47

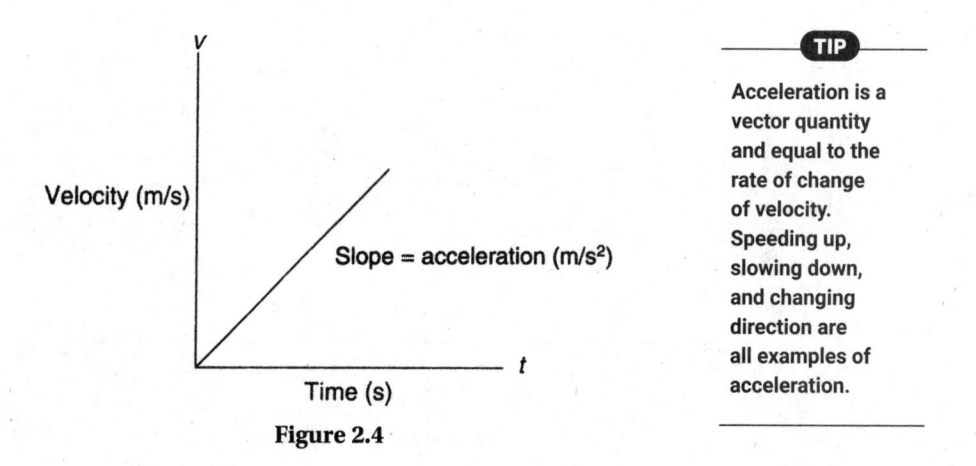

Figure 2.4

> **TIP**
> Acceleration is a vector quantity and equal to the rate of change of velocity. Speeding up, slowing down, and changing direction are all examples of acceleration.

The displacement from $t = 0$ to any other time is equal to the area of the triangle formed. However, between any two intermediary times, the resulting figure is a trapezoid. If we make several measurements, the displacement versus time graph for uniformly accelerated motion is a parabola starting from the origin (if we make the initial conditions that, when $t = 0$, $x = 0$, and $v = 0$; see Figure 2.5).

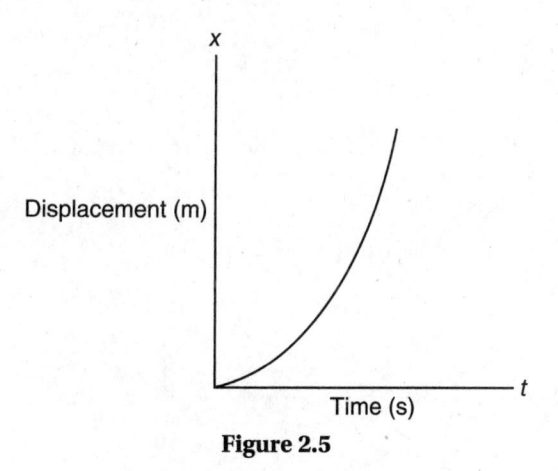

Figure 2.5

The slope of the velocity versus time graph is defined to be the average acceleration in units of meters per second squared (m/s²). We can now write, for the average acceleration (Figure 2.6),

$$\vec{a}_{\text{avg}} = \frac{\Delta \vec{v}}{\Delta t}$$

Figure 2.6

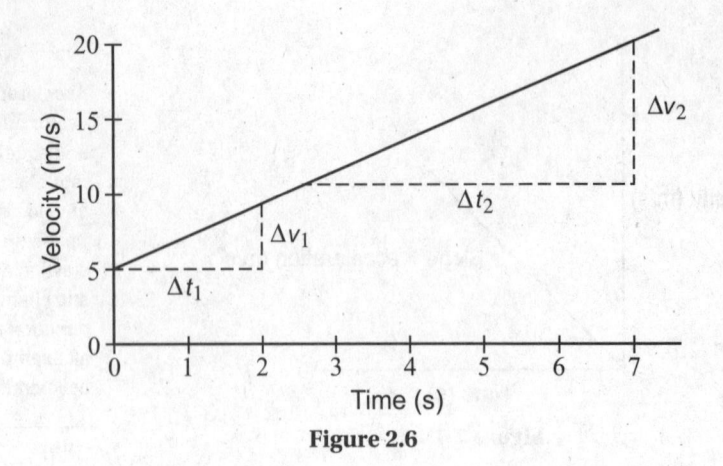

> **REMEMBER**
>
> An object with zero acceleration is not necessarily at rest in a given frame of reference. The object may be moving with constant velocity.

If the acceleration is taken to be constant in time, our expression for average acceleration can be written in a form that allows us to calculate the instantaneous final velocity after a period of acceleration has taken place. In other words, $\Delta \vec{v} = \vec{a}t$ (if we start our time interval from zero). If we define $\Delta \vec{v}$ to be equal to the difference between a final and an initial velocity $\vec{v}_f$ and $\vec{v}_i$ respectively, we can arrive at the fact that

$$\vec{v}_f = \vec{v}_i + \vec{a}t$$

If we plot velocity versus time for uniformly accelerated motion starting with a nonzero initial velocity, we get a graph that looks like Figure 2.7.

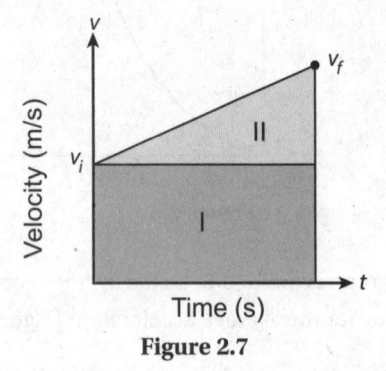

Figure 2.7

The displacement during any period of time will be equal to the total area under the graph. In this case, the total area will be the sum of two areas, one a triangle and the other a rectangle. The area of the rectangle, for some time t, is just $v_i t$. The area of the triangle is one-half the base times the height. The "base" in this case is the time period, t; the "height" is the change in velocity, Δv. Therefore, the area of the triangle is $\tfrac{1}{2}\Delta v t$. If we recall the definition of Δv and the fact that $v_f = v_i + at$, we obtain the following formula for the displacement during uniformly accelerated motion starting with an initial velocity:

$$x = v_i t + \frac{1}{2}at^2$$

In order to take into account a nonzero starting position (x_0), this formula can also be written as

$$x = x_0 + v_i t + \frac{1}{2}at^2$$

This analysis suggests an alternative method of determining the average velocity of an object during uniformly accelerated motion. The area under the flat line halfway between v_i and v_f will be the same as in Figure 2.7.

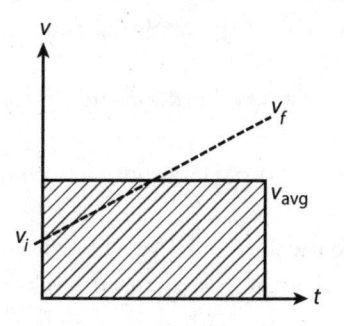

Therefore, we can simply write

$$v = \frac{v_i + v_f}{2} = v_{avg}$$

Since the above equation relates the average velocity to the initial and final velocities (for uniformly accelerated motion), we can write our displacement formula as

$$x = \frac{v_i + v_f}{2} t = (v_{avg}) t$$

Occasionally, a problem in kinematics does not explicitly mention the time involved. For this reason, it would be nice to have a formula for velocity that does not involve the time factor. We can derive one from all the other formulas.

Now, since $v_f = v_i + at$, we can express the time as $t = (v_f - v_i)/a$. Therefore

$$x = \left(\frac{v_i + v_f}{2}\right)\left(\frac{v_f - v_i}{a}\right)$$

$$v_f^2 - v_i^2 = 2ax$$

> **Sample Problem**

Motion can be shown graphically. Identify each region of the following graph as having negative, positive, or zero average acceleration for that interval of time.

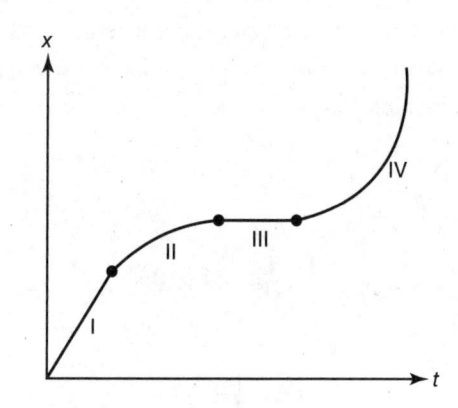

Solution

The instantaneous slopes on the given position versus time graph are the velocity values. To determine the acceleration, we must look at how those slopes are changing over the particular interval.

   I. All instantaneous slopes have the same value during this interval. There is no change in velocity: **zero acceleration**.
   II. Instantaneous slopes are decreasing (from large positive to zero) during this interval: **negative acceleration**.
   III. No change in slope equals **zero acceleration**. (Note that the velocity also happens to be zero here.)
   IV. Instantaneous slopes are increasing (from small positive values to large ones): **positive acceleration**.

Sample Problem

A particle accelerates from rest at a uniform rate of 3 m/s² for a distance of 200 m. How fast is the particle going at that time? How long did it take for the particle to reach that velocity?

Solution

We use the formula

$$v_f^2 - v_i^2 = 2ax$$

Since the particle begins from rest, the initial velocity is equal to zero and

$$v_f^2 = 2(3 \text{ m/s}^2)(200 \text{ m})$$

$$v_f = 34.64 \text{ m/s}$$

To find the time, we use $v = at$

$$t = \frac{34.64 \text{ m/s}}{3 \text{ m/s}^2} = 11.55 \text{ s}$$

Since velocity and acceleration are vector quantities, we need to consider the algebraic conventions accepted for dealing with various directions. For example, we usually agree to consider motion up or to the right as positive, and motion down or to the left as negative. Note that negative velocity means motion to the left (or down) and does not imply the object is slowing down. Likewise, negative acceleration means velocity is changing in the negative direction and does not imply the object is slowing down. When people say an object is decelerating, however, they do mean the object is slowing down. That can happen in two different ways. First, an object with positive velocity can slow down with a negative acceleration. Second, an object with a negative velocity can slow down with a positive acceleration. The general rule is that if the velocity and acceleration are pointed in the same direction, the object speeds up. However, if they are in opposite directions, the object slows down. (Acceleration at a right angle to velocity makes an object change direction.)

Examples

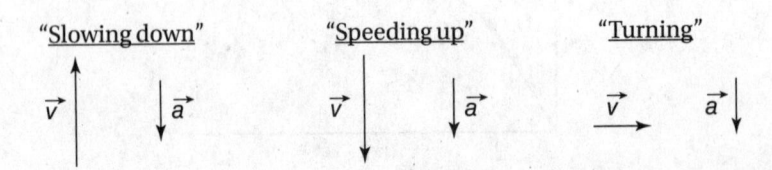

"Slowing down"   "Speeding up"   "Turning"

Recall that acceleration is rate of change of velocity, so the direction of the acceleration can always be found from doing a quick head-to-tail graphical solution of $\vec{v}_f - \vec{v}_i$. For example, when undergoing circular (or any turning) motion, the acceleration can be shown to be inward:

Accelerated Motion Due to Gravity

Gravity provides a constant downward acceleration. The value of gravity on Earth is commonly represented by the symbol g to stand in for 9.8 m/s/s. Note that an object going upward will be slowed down by gravity, while a downward moving object is sped up by gravity. A sideways-moving object will turn toward the ground! An object undergoing acceleration due to gravity and gravity alone (with no other forces) is referred to as being in "free fall." This is true regardless of whether the object is moving vertically up or down. Free fall is also called "projectile motion" if the object is also moving sideways while under the influence of gravity.

Choosing up as positive and down as negative for all three kinematic vectors (displacement, velocity, and acceleration) will provide consistent results for vertical velocities and vertical displacements:

$$v_f = v_i + at \quad \text{becomes} \quad v_f = v_i - gt \quad \text{since } a = -g$$

Using Δy to represent the vertical displacement:

$$\Delta y = v_i + \frac{1}{2}at^2 \quad \text{becomes} \quad \Delta y = v_i - \frac{1}{2}gt^2$$

Note that if an object is dropped or falls, $v_i = 0$. All subsequent displacements and velocities are negative since all motion is downward in those cases. If an object is thrown upward, v_i is positive. If an object is thrown downward, v_i is negative.

> **REMEMBER**
> Students often forget that there is still gravitational acceleration even at the top of a projectile's path! The acceleration vector never changes during free fall (including during projectile motion).

▶ Sample Problem

A projectile is fired vertically upward at an initial velocity of +98 m/s. How high will it rise? How long will it take the projectile to reach that height?

> **TIP**
> AP Physics 1 allows you to approximate Earth's gravity as 10 m/s² if you prefer to simply the calculations.

✓ Solution

We use $a = -9.8$ m/s² (recall gravity's direction is downward) and note that $v_{fy} = 0$ at the highest point:

$$v_{fy}^2 - v_{iy}^2 = 2ay$$
$$-(98 \text{ m/s})^2 = -2(9.8 \text{ m/s}^2)y_{max}$$
$$y_{max} = 490 \text{ m}$$

To find the time, we use

$$v = at$$
$$t = \frac{-98 \text{ m/s}}{-9.8 \text{ m/s}^2} = 10 \text{ s}$$

Graphical Analysis of Motion

We have seen that much information can be obtained if we consider the graphical analysis of motion. If complex changes in motion are taking place, visualization may provide a better understanding of the physics involved than algebra. The techniques of graphical analysis are as simple as slopes and areas. For example, we already know that, for uniformly accelerated motion, the graph of distance versus time is a parabola. Since the slope is changing, the instantaneous velocity can be approximated at a point P by finding the slope of a tangent line drawn to a given point on the curve (see Figure 2.8).

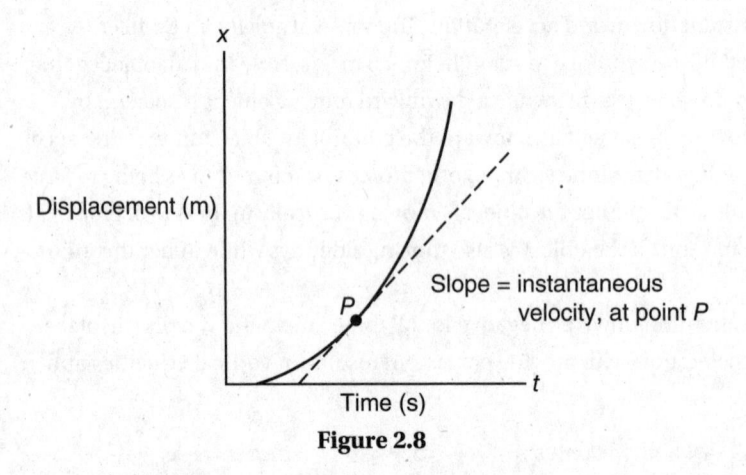

Figure 2.8

What would happen if an object accelerated from rest, maintained a constant velocity for a while, and then slowed down to a stop? Using what we know about graphs of velocity and acceleration in displacement versus time, we might represent the motion as shown in Figure 2.9.

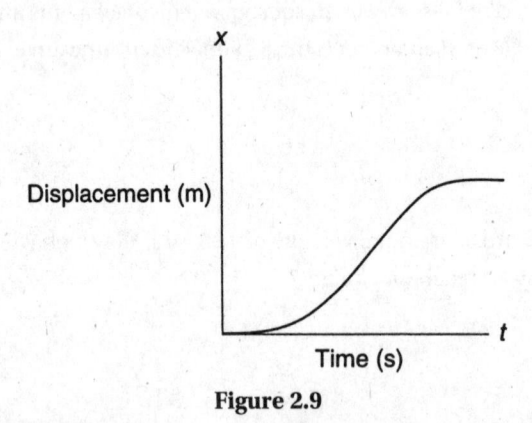

Figure 2.9

We can apply many instances of motion to graphs. In the case of changing velocity, consider the graph of velocity versus time for an object thrown upward into the air, reaching its highest point, changing direction, and then accelerating downward. This motion has a constant downward acceleration that, at first, acts to slow the object down, but later acts to speed it up. A graph of this motion is seen in Figure 2.10.

(a)

Graph showing height vs. time as a parabola with "Highest point" labeled at the peak and "slope = velocity = changing" noted on the curve.

(b)

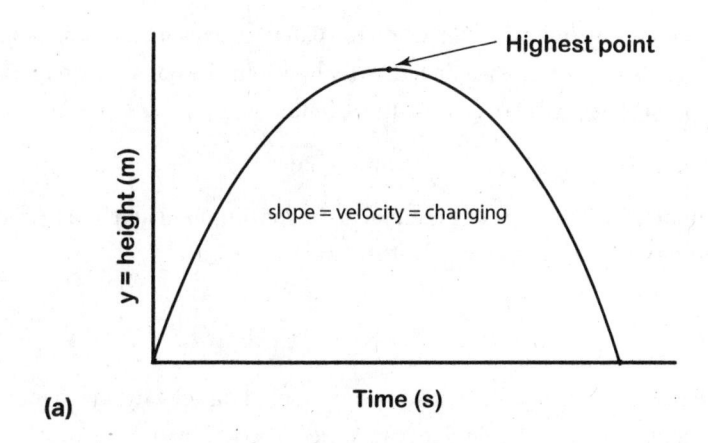

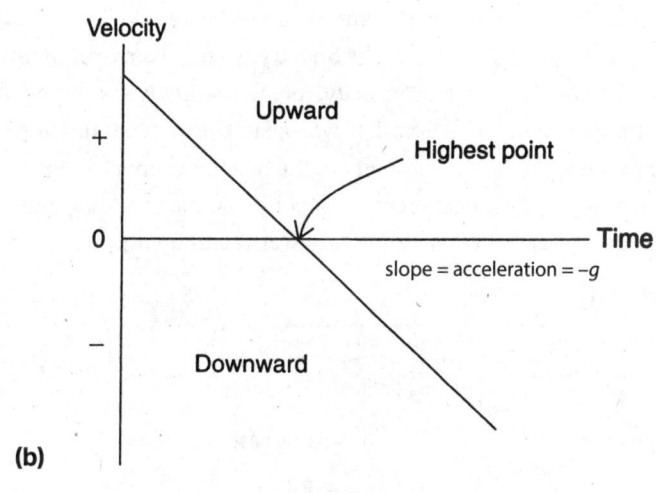

Graph showing velocity vs. time as a decreasing straight line passing through zero at the "Highest point", with "Upward" above the axis (+) and "Downward" below (−), and "slope = acceleration = −g".

Figure 2.10

Figure 2.10a shows the parabolic nature of $\Delta y = v_i t - \frac{1}{2}gt^2$.
Figure 2.10b shows the linear nature of $v_f = v_i - gt$.

▶ Sample Problem

A ball is thrown straight up into the air and, after being in the air for 9 s, is caught by a person 5 m above the ground. To what maximum height did the ball go?

✓ Solution

The equation

$$y = v_i t - \frac{1}{2}gt^2$$

represents the vertical position of the ball above the ground for any time t. Since the ball is thrown upward, the initial velocity is positive, while the acceleration of gravity is always directed downward (and hence is algebraically negative). Thus, we can use the above equation to find the initial velocity when $y = 5$ m and $t = 9$ s. Substituting these values, as well as the magnitude of g (9.8 m/s^2) in the equation, we get $v_i = 44.65$ m/s.

Now, to find the maximum height, we note that, when the ball rises to its maximum, its speed becomes zero since its velocity is changing direction. Since we do not know how long the ball takes to rise to its maximum height (we could determine this value if desired), we can use the formula

$$v_f^2 - v_i^2 = -2gy$$

When the final velocity equals zero, the value of y is equal to the maximum height. Using our answer for the initial velocity and the known value of g, we find that $y_{max} = 101.7$ m.

Relative Motion

All measurements are relative. Specifically, all measurements, including velocity, are made with respect to some reference point or object. Just as vectors are all relative to a specific coordinate system, all velocities are too. When we describe a car as moving at 55 mph, what we really mean is 55 mph relative to Earth's surface. The rules of vector addition provide us with a means by which to translate one relative velocity to another.

An example of this type of motion can be seen when a boat trying to cross a river or an airplane meeting a crosswind is considered. In the case of the boat, its velocity, relative to the river, is based on the properties of the engine and is measured by the speedometer on board. However, to a person on the shore, its **relative velocity** (or effective velocity) is different from what the speedometer in the boat may report. In Figure 2.11, we see such a situation with the river moving to the right at 4 meters per second and the boat moving relative to the water at 10 meters per second. If you like, we can call to the right "eastward" and up the page "northward."

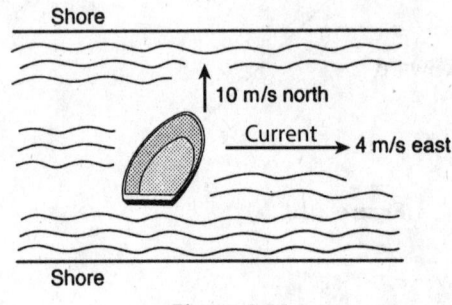

Figure 2.11

> **TIP**
> Although students should understand two-dimensional relative velocity problems like this one, they will only be asked to quantitatively calculate one-dimensional relative velocity problems.

By vector methods, the resultant velocity relative to the shore is given by the Pythagorean theorem. The direction is found by means of a simple sketch (Figure 2.12) connecting the vectors head to tail to preserve the proper orientation. Numerically, we can use trigonometry.

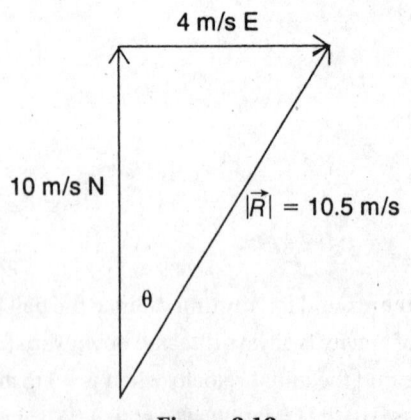

Figure 2.12

The resultant velocity is 10.5 meters per second at an angle of 22 degrees east of north.

Horizontally Launched Projectiles

If you roll a ball off a smooth table, you will observe that it does not fall straight down. With trial and error, you might observe that how far it falls will depend on how fast it is moving forward. Initially, however, the ball has no vertical velocity. The ability to "fall" is given by gravity, and the acceleration due to gravity is 9.8 meters per second squared downward. Since gravity acts vertically and the initial velocity (v_i) is horizontal, the two motions are simultaneous and independent. Galileo demonstrated that the trajectory characterized by constant horizontal velocity and constant vertical acceleration is a parabola.

We know that the distance fallen (making the downward direction positive in this case) by a mass dropped from rest is given by the equation

$$\Delta y = v_{iy} t + \tfrac{1}{2} a_y t^2 = \tfrac{1}{2} g t^2$$

since $v_{iy} = 0$ and $a_y = g$

Since the ball that rolled off the table is moving horizontally, with some initial constant velocity, it covers a distance, x, called the **range**:

$$\Delta x = v_{ix} t + \tfrac{1}{2} a_x t^2 = v_{ix} t$$

since $a_x = 0$

Since the time is the same for both motions, we can first solve for the time, using the x equation, and then substitute it into the y equation. In other words, y as a function of x is easily shown to be

$$\Delta y = \tfrac{1}{2} g t^2 = \tfrac{1}{2} g (\Delta x / v_{ix})^2 = g \Delta x^2 / (2 v_{ix}^2)$$

which is of course the equation of a parabola. This equation of y in terms of x is called the **trajectory** of the projectile, while the two separate equations for x and y as functions of time are called **parametric equations**. Figure 2.13 illustrates this trajectory as well as a position vector $\vec{R}$, which locates a point at any given time in space.

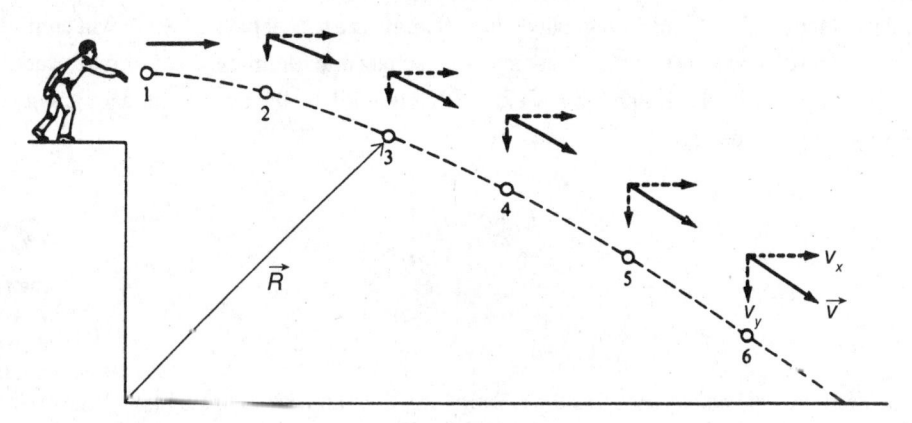

Figure 2.13

If the height from which a projectile is launched is known, the time to fall can be calculated from the equation for free fall. For example, if the height is 49 m, the time to fall is 3.16 s. If the horizontal velocity is 10 m/s, the maximum range will be 31.6 m.

$$\Delta y = \tfrac{1}{2} g t^2$$
$$49 = \tfrac{1}{2}(9.8) t^2$$
$$t = 3.2 \text{ s}$$
$$\Delta x = v_{ix} t = 10(3.2) = 32 \text{ m}$$

> **REMEMBER**
>
> For all projectiles, the two component motions are independent of each other. The horizontal velocity remains constant throughout the projectile's motion.

Sample Problem

A projectile is launched horizontally from a height of 25 m, and it is observed to land 50 m from the base. What was the launch velocity?

Solution

We know that the vertical motion is independent of the horizontal motion. Thus, we can find the time that the projectile is in the air using:

$$d = \frac{1}{2}at^2$$

Solving for t we get

$$t = 2.26 \text{ s}$$

Now, since $x = v_x t$, we see that $v_x = \frac{50 \text{ m}}{2.26 \text{ s}} = 22.12$ m/s.

Projectiles Launched at an Angle

Suppose that a rocket on the ground is launched with some initial speed, v, at some angle θ. The vector nature of velocity allows us to immediately write the equations for the horizontal and vertical components of initial velocity:

$$v_{ix} = v \cos\theta \quad \text{and} \quad v_{iy} = v \sin\theta$$

Since each motion is independent, we can consider the fact that, in the absence of friction, the horizontal velocity will be constant while the y velocity will decrease as the rocket rises. When the rocket reaches its maximum height, its vertical velocity will be zero; then gravity will accelerate the rocket back down. It will continue to move forward at a constant rate. How long will the rocket take to reach its maximum height? From the definition of acceleration and the equations earlier in this chapter, we know that this will be the time needed for gravity to decelerate the vertical velocity to zero—that is,

$$v_y = v_{iy} + at$$
$$0 = v_{iy} - g t_{up}$$
$$t_{up} = \frac{v_{iy}}{g}$$

The total time of flight will be just twice this time. Therefore, the range is the product of the initial horizontal velocity (which is constant) and the total time. In other words,

$$R = \text{range} = v_{ix}(2 t_{up}) = 2v \cos\theta \, t_{up} = 2v \cos\theta \, (v \sin\theta / g)$$
$$= 2v^2 \cos\theta \sin\theta / g$$

Since the vertical motion is independent of the horizontal motion, the changes in vertical height are given one-dimensionally as

$$\Delta y = v_{iy} t - \frac{1}{2}gt^2$$

substituting in for v_{iy} its general expression in term speed and angle:

$$\Delta y = (v \sin\theta) t - \frac{1}{2}gt^2$$

> **REMEMBER**
>
> When a projectile is launched at an angle, the vertical velocity component is equal to zero at the maximum height. Additionally, for a given launch speed, the maximum range occurs when the launch angle is equal to 45° (when launching and landing at the same height).

If we want to know the maximum height achieved, we simply use the value for the time to reach the highest point. The trajectory is seen in Figure 2.14 for a baseball being hit.

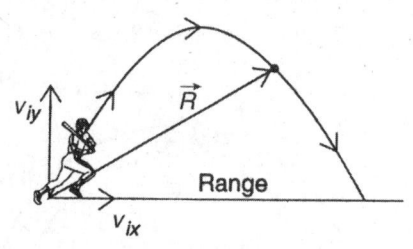

Figure 2.14

As an example, if a projectile is launched with an initial velocity of 100 m/s at an angle of 30°, the maximum range will be equal to 883.7 m. To find the maximum height, we could first find the time required to reach that height, but a little algebra will give us a formula for the maximum height independent of time. The maximum height can also be expressed as

$$\Delta y = v_{iy} t + \frac{1}{2} a t^2$$
$$\Delta y_{max} = v_{iy} t_{up} - \frac{1}{2} g (t_{up})^2$$
$$= v_{iy} \left(\frac{v_{iy}}{g}\right) - \frac{1}{2} g \left(\frac{v_{iy}}{g}\right)^2$$
$$= \frac{v_{iy}^2}{2g}$$

Using the known numbers, we find that the maximum height reached is 127.55 m.

> **Sample Problem**

A projectile is launched from the ground at a 40° angle with a velocity of 150 m/s. Calculate the maximum height of the projectile.

> **Solution**

From trigonometry, we know that $v_{iy} = v_0 \sin \theta$. Thus,

$$v_{iy} = (150 \text{ m/s}) \sin(40°) = 96 \text{ m/s}$$

Now, at the maximum height, the vertical velocity is equal to zero, and the acceleration is g:

$$v_{fy}^2 - v_{iy}^2 = -2gy$$
$$y_{max} = \frac{v_{iy}^2}{2g} = \frac{96 \text{ m}^2/\text{s}^2}{19.6 \text{ m/s}^2} = 474 \text{ m}$$

Uniform Circular Motion

Velocity is a vector. To change a velocity's direction simply requires an acceleration since change in direction is a change in the vector. An acceleration directed perpendicularly to the velocity will change the direction of the velocity in the direction of the acceleration. As shown on page 51, when the direction is the only quantity changing, as the result of a centrally directed deflecting force, the result is uniform circular motion. We consider here an object already undergoing periodic, uniform circular motion. By this description we mean that the object maintains a constant speed as it revolves around a circle of radius r, in a period of time T. The number of revolutions per second is called the **frequency**, f. This is illustrated in Figure 2.15.

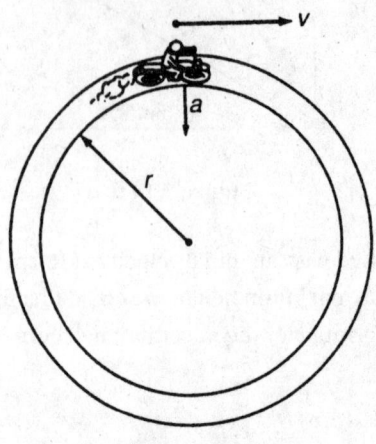

Figure 2.15

The direction of the acceleration is toward the center and is called the **centripetal acceleration**. The magnitude of the centripetal acceleration is given by two formulas:

$$a_c = \frac{v^2}{r}$$

Recalling that the circumference of a circle is $2\pi r$, the speed must be:

$$v = \frac{2\pi r}{T} \text{ for uniform circular motion}$$

Substituting this expression for the tangential velocity into our formula for centripetal acceleration:

$$a_c = \frac{4\pi^2 r}{T^2}$$

Oftentimes, circular motion (or rotational motion in general) is described in terms of period or frequency of rotation rather than speed:

$$T = \frac{1}{f} \text{ (relationship between period and frequency)}$$

Sometimes, it is more convenient to describe rotational motion in terms of angular frequency (radians/sec rather than cycles/sec). Angular frequency is also the rate of change of the angular position and so is also known as the rotational (or angular) velocity. Since there are 2π radians in one complete cycle, the relationship between angular frequency (ω) and frequency is a simple one:

$$\omega = 2\pi f$$

> **Sample Problem**

A 5-kg mass is undergoing uniform circular motion with a constant speed of 10 m/s in a circle of radius 2 m. Calculate the centripetal acceleration of the mass.

✓ **Solution**

We use the formula

$$a_c = \frac{v^2}{r} = \frac{(10 \text{ m/s})^2}{2 \text{ m}} = 50 \text{ m/s}^2$$

SUMMARY

- Displacement, velocity, and acceleration are all vectors.
- Distance, speed, and time are scalars.
- Kinematics is a description of motion.
- Velocity is defined to be equal to the rate of change of displacement.
- Acceleration is defined to be equal to the rate of change of velocity.
- Velocity is the slope of a displacement versus time graph.
- Acceleration is the slope of a velocity versus time graph.
- The displacement can be obtained from a velocity versus time graph by taking the area under the graph.
- The acceleration due to gravity ($\vec{g}$) near the surface of Earth is equal to 9.8 m/s² and is directed downward. This is the only acceleration in free fall and projectile motion problems.
- The relative velocity can be found by vector addition of the individual velocity vectors.
- When a projectile is launched, in the absence of air resistance, the horizontal motion is independent of the vertical motion.
- The vertical and horizontal components of the launch velocity can be obtained using trigonometry. The regular kinematics equations can then be used for each direction.
- If an object is moving in a circle, then the object has an acceleration acting toward the center called the centripetal acceleration.

Problem-Solving Strategies for Motion Problems

Since all motion is relative, it is very important to always ask yourself the following question: "From what frame of reference am I viewing the situation?" Also, whenever you solve a physics problem, be sure to consider the assumptions being made, explicitly or implicitly, about the moving object or objects. In this way, your ability to keep track of what is relevant for your solution path will be maintained. In addition, it is suggested that you:

1. Identify all the goals and givens in the problem. Recall from the Introduction that the goals may be explicit or implicit. If a question is based on a decision or prediction, be sure to understand the algebraic requirements necessary to reach an answer.
2. Consider the meaningfulness of your solution or process. Does your answer or methodology make sense?
3. Choose a proper coordinate system and remember the sign conventions for algebraically treating vector quantities. In addition, make sure you understand the nature of the concepts being discussed (for example, whether they are vectors or scalars).

4. Use proper SI units throughout your calculations, and make sure that correct units are included in your final answer. Try dimensional analysis to see whether the answer makes sense. Perhaps it looks too large or too small because it is expressed in the wrong units.
5. Make a sketch of the situation if one is not provided.
6. If you are interpreting a graph, be sure you understand the interrelationships among all the kinematic variables discussed (slopes, areas, etc.).
7. If you are constructing a graph, label both axes completely, choose a proper scale for each axis, and draw neatly and clearly.
8. Try different problem-solving heuristics if you get stuck on a difficult problem.
9. Remember to break vectors into components first if the problem is two-dimensional.

Practice Exercises

Multiple-Choice

1. A ball is thrown upward with an initial velocity of 20 m/s. How long will the ball take to reach its maximum height?

    (A) 19.6 s
    (B) 9.8 s
    (C) 6.3 s
    (D) 2.04 s

2. An airplane lands on a runway with a velocity of 150 m/s. How far will it travel until it stops if its rate of deceleration is constant at -3 m/s^2?

    (A) 525 m
    (B) 3,750 m
    (C) 6,235 m
    (D) 9,813 m

3. A ball is thrown downward from the top of a roof with a speed of 25 m/s. After 2 s, its velocity will be

    (A) 19.6 m/s
    (B) -5.4 m/s
    (C) -44.6 m/s
    (D) 44.6 m/s

4. A rocket is propelled upward with an acceleration of 25 m/s^2 for 5 s. After that time, the engine is shut off, and the rocket continues to move upward. The maximum height, in meters, that the rocket will reach is

    (A) 900
    (B) 1,000
    (C) 1,100
    (D) 1,200

Questions 5–7 refer to the velocity versus time graph shown below.

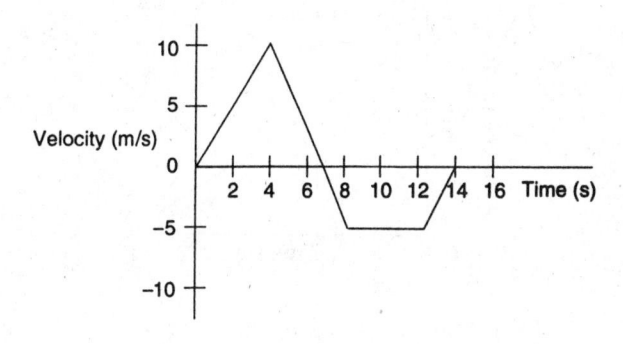

5. The total distance traveled by the object during the indicated 14 s is

    (A) 7.5 m
    (B) 25 m
    (C) 62.5 m
    (D) 77.5 m

6. The total displacement of the object during the 14 s indicated is

    (A) 7.5 m
    (B) 25 m
    (C) 62.5 m
    (D) 77.5 m

7. The average velocity over the whole time, in meters per second, of the object is

    (A) 0
    (B) 0.5
    (C) 2.5
    (D) 4.5

8. What is the total change in velocity for the object whose acceleration versus time graph is given below?

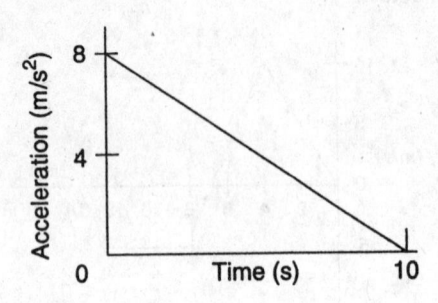

[Graph: Acceleration (m/s²) vs Time (s); line from (0, 8) to (10, 0)]

(A) 40 m/s
(B) −40 m/s
(C) 80 m/s
(D) −80 m/s

9. An object has an initial velocity of 15 m/s. How long must it accelerate at a constant rate of 3 m/s² before its average velocity is equal to twice its initial velocity?

(A) 5 s
(B) 10 s
(C) 15 s
(D) 20 s

10. A projectile is launched at an angle of 45° with a velocity of 250 m/s. If air resistance is neglected, the magnitude of the horizontal velocity of the projectile at the time it reaches maximum altitude is equal to

(A) 0 m/s
(B) 175 m/s
(C) 200 m/s
(D) 250 m/s

11. A projectile is launched horizontally with a velocity of 25 m/s from the top of a 75-m height. How many seconds will the projectile take to reach the bottom?

(A) 15.5
(B) 9.75
(C) 6.31
(D) 3.91

12. At a launch angle of 45°, the range of a launched projectile is given by

(A) $\dfrac{v_i^2}{g}$

(B) $\dfrac{2v_i^2}{g}$

(C) $\dfrac{v_i^2}{2g}$

(D) $\sqrt{\dfrac{v_i^2}{2g}}$

13. A projectile is launched at a certain angle. After 4 s, it hits the top of a building 500 m away. The height of the building is 50 m. The projectile was launched at an angle of

(A) 14°
(B) 21°
(C) 37°
(D) 76°

14. The operator of a boat wishes to cross a 5-km-wide river that is flowing to the east at 10 m/s. He wants to reach the exact point on the opposite shore 15 min after starting. With what speed and in what direction should the boat travel?

(A) 11.2 m/s at 26.6° E of N
(B) 8.66 m/s at 63.4° W of N
(C) 11.4 m/s at 60.9° W of N
(D) 8.66 m/s at 26.6° E of N

Additional Practice

1. A particle has the acceleration versus time graph shown below:

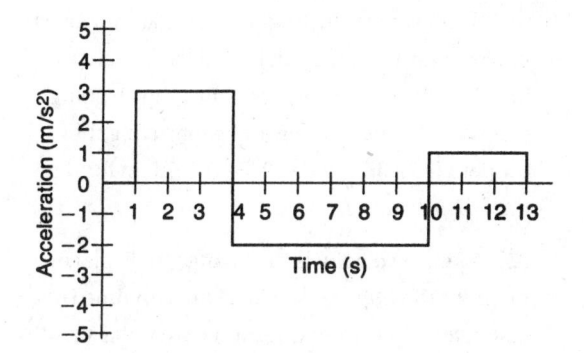

   [Graph: Acceleration (m/s²) vs Time (s). Acceleration is 3 m/s² from t=1 to t=4, then -2 m/s² from t=4 to t=10, then 1 m/s² from t=10 to t=13.]

   If the particle begins its motion at $t = 0$, with $v = 0$ and $x = 0$, make a graph of velocity versus time.

2. A stone is dropped from a 75-m-high building. When this stone has dropped 15 m, a second stone is thrown downward with an initial velocity such that the two stones hit the ground at the same time. What was the initial velocity of the second stone?

3. A particle is moving in one dimension along the x-axis. The position of the particle is given, for any time, by the following position function:

   $$x(t) = 3t^2 + 2$$

   (a) Evaluate the average velocity of the particle starting at $t = 2$ s for $\Delta t = 1, 0.5, 0.2, 0.1, 0.01, 0.001$ s.
   (b) Interpret the physical meaning of your results for part (a).

4. A stone is dropped from a height h and falls the last half of its distance in 4 seconds. (a) What was the total time of fall? (b) From what height was the stone dropped?

5. If the average velocity of an object is nonzero, does this mean that the instantaneous velocity of the object can never be zero? Explain.

6. A girl standing on top of a roof throws a stone straight upward into the air with a certain speed. She then throws a similar stone directly downward with the same speed. Compare the velocities of both stones when they reach the ground.

7. Explain how it is possible for an object to have zero average velocity and still have nonzero average speed.

8. A mass attached to a string is twirled overhead in a horizontal circle of radius r every 0.3 s. The path is directly 0.5 m above the ground. When released, the mass lands 2.6 m away. What was the velocity of the mass when it was released, and what was the radius of the circular path?

9. A football quarterback throws a pass to a receiver at an angle of 25 degrees to the horizontal and at an initial velocity of 25 m/s. The receiver is initially at rest 30 m from the quarterback. The instant the ball is thrown, the receiver runs at a constant velocity to catch the pass. In what direction and at what speed should he run?

10. A car is moving in a straight line with velocity v. Raindrops are falling vertically downward with a constant terminal velocity u. At what angle does the driver think the drops are hitting the car's windshield? Explain.

Answers Explained

Multiple-Choice Problems

1. **(D)** Given the initial velocity of 20 m/s, we know that the ball is decelerated by gravity at a rate of 9.8 m/s². Therefore, we need to know how long gravity will take to decelerate the ball to zero velocity. Clearly, from the definition of acceleration, dividing 20 m/s by 9.8 m/s² gives the answer: 2.04 s.

2. **(B)** We do not know the time needed to stop, but we do know that the final velocity is zero. If we use the formula
$$v_f^2 - v_t^2 = 2ax$$
and substitute 150 m/s for the initial velocity and −3 m/s² for the deceleration, we get the answer: 3,750 m.

3. **(C)** The initial velocity is −25 m/s downward, so $v = -(25) - (9.8)(2) = -44.6$ m/s.

4. **(C)** The rocket accelerates from rest, so the first distance traveled in 5 s is given by
$$y_1 = \frac{1}{2}at^2$$
Substituting the values for time (5 s) and acceleration (25 m/s²) gives $y_1 = 312.5$ m. After 5 s of accelerating, the rocket has a velocity of 125 m/s. After this time, the engine stops and the rocket is decelerated by gravity as it continues to move upward. The time to decelerate to zero is found by dividing 125 m/s by the acceleration of gravity. Thus, 125/9.8 = 12.76 s.

    The distance traveled during that time is added to the first accelerated distance. The second distance is given by
$$y_2 = 125(12.76) - 4.9(12.76)^2 = 796.6 \text{ m}$$
The total distance traveled is therefore equal to 1,108 m or, rounded off, 1,100 m.

5. **(C)** The total distance traveled by the object in 14 s is equal to the total area under the graph. Breaking the figure up into triangles and rectangles, we find that their areas add up to 62.5 m (recall that distance is a scalar).

6. **(A)** The total displacement for the object is the sum of the positive (forward) and negative (backward) areas, representing the fact that the object moves away from the origin and then back (as indicated by the velocity vs. time graph), going below the x-axis. Adding and subtracting the proper areas: 35 − 2.5 − 20 − 5 = 7.5 m for the final displacement.

7. **(B)** Average velocity is the change in displacement for the object over the change in time (not distance). Since from question 6 we know the displacement change to be 7.5 m in 14 s, the average velocity is therefore 0.53 m/s.

8. **(A)** The change in velocity for the object is equal to the area under the graph, which is equal to 1/2 (+8)(10) = +40 m/s.

9. **(B)** For uniformly accelerated motion, the average velocity is
$$v = \frac{v_f + v_i}{2}$$
The question requires that the average velocity be equal to twice the initial velocity; thus, if the average velocity is 30 m/s, the final velocity attained must be 45 m/s. Now the question is, how long must the object accelerate from 15 m/s to achieve a speed of 45 m/s? A change in velocity of 30 m/s at a rate of 3 m/s² for 10 s is implied.

10. **(B)** The horizontal component of velocity remains constant in the absence of resistive forces and is equal to $v_i \cos \theta$. Substituting the known numbers, we get $v_x = (250)(0.7) = 175$ m/s.

11. **(D)** The time to fall is given by the free-fall formula:
$$t = \sqrt{\frac{2y}{g}}$$
If we substitute the known numbers, we get 3.91 s for the time.

12. **(A)** From the formula for range, we see that, at 45°, sin 2θ = 1, and so
$$R = \frac{v_i^2}{g}$$

13. **(A)** We know that, after 4 s, the projectile has traveled horizontally 500 m. Therefore, the horizontal velocity was a constant 125 m/s and is equal to $v_i \cos \theta$. We also know that, after 4 s, the y-position of the projectile is 50 m. Thus, we can write:

$$50 = 4 v_i \sin \theta - 4.9(4)^2 = 4 v_i \sin \theta - 78.4$$

Therefore, $v_i \cos \theta = 125$, $v_i \sin \theta = 32.1$, and $\tan \theta = 0.2568$, so $\theta = 14.4°$.

14. **(C)** The river is flowing at 10 m/s to the right (east), and the resultant desired velocity is 5.55 m/s up (north). Therefore, the actual velocity, relative to the river, is heading W of N. By the Pythagorean theorem, the velocity of the boat must be 11.4 m/s. The angle is given by the tangent function. In the diagram below, not drawn to scale but correct for orientation, $\tan(\theta) = \frac{10}{5.55} = 1.8$. Therefore, $\theta = 60.9°$ W of N.

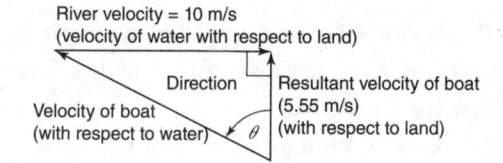

Additional Practice

1. The acceleration graph given in the problem indicates constant acceleration from $t = 1$ to $t = 3$. Therefore, the area represents a constant change in velocity from 0 to 9 m/s. The second region corresponds to negative acceleration, which slows the object down and turns it around. We know this since the area is -12 m/s. This brings the graph from 9 m/s to -3 m/s (a change in direction). The last area is a change of 3 m/s, which brings the velocity to 0 m/s. A graph of this motion is shown below, as desired:

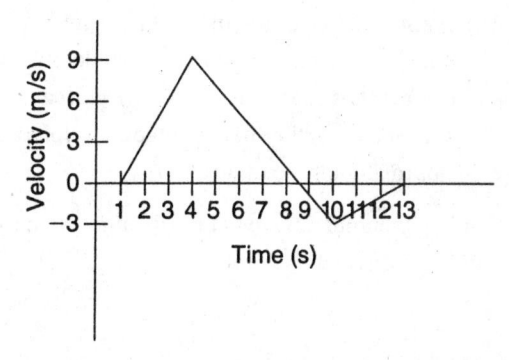

2. In this problem, we know that the first stone will be dropped from a height of 75 m, and it will be in free fall because of gravity. Therefore, the time to fall is given by

$$t = \sqrt{\frac{2y}{g}} = 3.9 \text{ s}$$

Since the first stone is to fall 15 m before the second stone is dropped, we can likewise determine that $t = 1.75$ s to fall that distance. Thus, since the two stones must reach the ground at the same time, we know that the second stone must reach the ground after the first one has traveled for 3.9 s. Since the second stone is thrown when the distance fallen by the first stone is 15 m, we subtract 1.75 s from the total time of 3.9 s to get the second stone's duration of fall. That time is equal to 2.15 s. Since the second stone is thrown downward, it has a negative displacement, and so we can write

$$-75 = -v_i(2.15) - \tfrac{1}{2}(9.8)(2.15)^2$$

This gives us a downward initial velocity of 24.34 m/s for the second stone.

3. (a) Using the position function, we know that when $t = 2$ s, $x = 14$ m. The average velocity is given by $\Delta x/\Delta t$. For $\Delta t = 1$ s, we find that at $t = 3$ s, $x = 29$ m. Thus, $\Delta x = 15$ m and the average velocity is equal to 15 m/s. If $\Delta t = 0.5$ s, then at $t = 2.5$ s, $x = 20.75$ m; the new average velocity is equal to 13.5 m/s (6.75 m/0.5 s). We continue this procedure. For $\Delta t = 0.2$ s, the average velocity is 12.6 m/s. For $\Delta t = 0.1$ s, the average velocity is 12.3 m/s. For $\Delta t = 0.01$ s, the average velocity is 12.03 m/s. Finally, for $\Delta t = 0.001$ s, the average velocity is 12.003 m/s.

(b) It appears that as the time interval gets smaller and smaller, the average velocity approaches 12 m/s as a limit. We can say that at $t = 2$ s, the "instantaneous" velocity is approximately equal to 12 m/s.

4. Let T = the total time to fall the distance h. In free fall, this is given by

$$h = \frac{1}{2}gT^2$$

Now we know that the stone falls the last half of its distance in 4 seconds.

This means

$$\frac{h}{2} = \frac{1}{2}g(T-4)^2$$
$$\frac{1}{4}gT^2 = \frac{1}{2}g(T-4)^2$$
$$T^2 = 2(T-4)^2$$
$$T = \sqrt{2}(T-4)$$

Solving for the total time yields $T = 13.7$ seconds, and, thus, $h = 915$ meters.

5. The average velocity is the ratio of the change in the displacement to the change in the time. It is possible that the object stops for a while and then continues. Thus, it is possible for a nonzero average velocity to have a zero instantaneous velocity.

6. The two stones will have the same speed when they reach the ground. As the first stone rises, gravity decelerates it until it stops, and then it begins to fall back down. When it passes its starting point, it has the same speed but in the opposite direction. This is the same starting speed as that of the second stone. Both stones are then accelerated through the same displacement, giving them both the same final velocity.

7. The average velocity is the total displacement divided by the total time and is a vector quantity. The average speed is equal to the total distance divided by the total time and is a scalar quantity. If an object is undergoing periodic motion, it can return to its starting point and have zero displacement in one period. This gives it zero average velocity. However, since it traveled a distance, it has a nonzero average speed.

8. The mass is undergoing uniform circular motion. Therefore, it has a constant speed. The velocity when it is released can be determined from the height of the mass and the horizontal range.

We know it falls vertically 0.5 m in 0.32 s. During this time, it moves horizontally 2.6 m. Thus, the magnitude of the horizontal initial velocity was

$$v = \frac{2.6 \text{ m}}{0.32 \text{ s}} = 8.125 \text{ m/s}$$

In circular motion, the constant velocity found above is equal to the ratio of the circumference of the circular path and the period:

$$v = \frac{2\pi r}{T}$$

Substituting for the velocity and the period of 0.3 s, we find that $r = 0.388$ m.

9. The first quantity we can calculate is the football's theoretical range:

$$R = \frac{v_i^2 \sin 2\theta}{g} = \frac{(25)^2 (\sin 50)}{g} = 48.85 \text{ m}$$

The receiver needs to travel 18.85 m away from the quarterback to catch the ball. To determine how fast the receiver must run, we need to know how long it takes the ball to travel 48.85 m horizontally. This is equal to the range divided by the horizontal velocity:

$$t = \frac{48.85 \text{ m}}{v_i \cos 25} = 2.16 \text{ s}$$

Therefore, to travel 18.85 m in 2.16 s, the receiver must run at 8.73 m/s.

10. From the diagram below, we can see that, to the driver, $\tan \theta = u/v$.

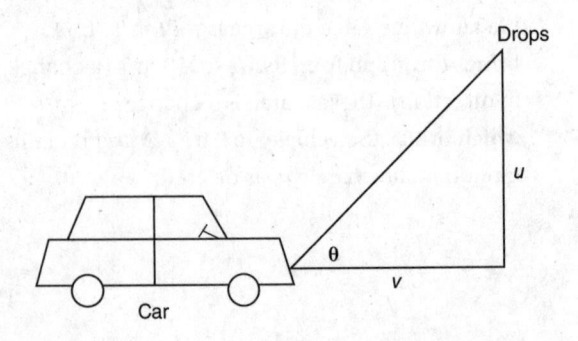

3

Forces and Newton's Laws of Motion

Learning Objectives

In this chapter, you will learn about:

- → Forces
- → Newton's laws of motion
- → Static applications of Newton's laws
- → Free-body diagrams
- → Dynamic applications of Newton's laws
- → Central forces
- → Friction

From our discussion in Chapter 1 of frames of reference, you should be able to convince yourself that, if two objects have the same velocity, the relative velocity between them is zero, and therefore one object looks as though it is at rest with respect to the other. In fact, it would be impossible to decide whether or not such an "**inertial**" observer was moving! Therefore, accepting this fact, we state that an object that appears to be at rest in the Earth frame of reference will simply be stated to be "at rest" relative to us (the observers).

Forces

An object does not change its velocity (accelerate) on its own. Rather, an interaction is required between that object and another object. This interaction can be direct as in the case of contact force, or it may take place via an action at a distance. Gravitational, electrical, and magnetic forces are all examples of forces that act at a distance (no direct contact is required). Friction and normal forces are examples of contact forces. If there is a net unbalanced force (the forces on an object do not add up to zero), then there will be an acceleration. This acceleration can be a change in either speed or direction of motion of the object, or it can be both. If there is no change in speed or direction, the object is not accelerating and the net force must be zero.

> **TIP**
> Forces are vector quantities that represent a push or a pull on an object.

Unless an object is in deep space, far from the influence of any other objects, objects experience a net force of zero because two or more forces are in opposition to each other, canceling out each other. These objects experiencing a net force of zero are in a force equilibrium. If the object is at rest, it is in **static equilibrium**. Note that an object can be in motion while experiencing a net force of zero; it may be experiencing a constant velocity.

The following is a complete list of all common forces used on the AP Physics 1 exam.

Types of Forces

Force	When Force Is Present	Rule for Direction
Weight/gravity	Almost always (the object could be *very* far from any planet)	Drawn straight toward the center of Earth
Normal	When an object is in contact with another object	Drawn straight away from the surface (perpendicular)
Tension	When an object is being pulled by a rope, string, or chain	Drawn in the direction the rope is attached
Push or pull	When a person or an active agent is interacting with the object	Made clear by the situation
Spring/restorative	When a spring or an elastic material is compressed or extended	Opposite the extension or compression
Friction	Surface friction: When two surfaces are sliding (kinetic) or trying to slide (static) past each other. Fluid friction: When an object is moving through a gas or liquid	Opposes the motion (velocity vector) or the likely motion if stationary

Of these forces, weight (or the force due to gravity when on the surface of a planet) is the most ubiquitous and has its own formula:

$$F_g = mg$$

In this equation, g is the strength of the gravity field at the location of the mass (9.8 N/kg at Earth's surface). Note that weight is a vector force and depends on an object's location in the universe. In contrast, mass is a scalar measure of an object's inertia, which belongs to the object independent of location. This g identified here as the gravitational field strength is the same g used in kinematics as the acceleration due to gravity in free fall and projectile motion.

When generating a list of everyday forces such as the one above, it is not always clear what to call a particular force. For example, if you push something with your hand, it could be called a normal force between two surfaces or it could be called simply a "push." These so-called contact forces (that is, all the forces in the table above except for gravity and electricity) are actually not fundamental in the sense that we are naming some overall effect that derives from many microscopic fundamental interactions. All forces on this list, except for gravity, are actually electromagnetic in nature. They are electromagnetic in the sense that they can ultimately be explained in terms of charges interacting with other charges. In fact, all of the forces you experience in your daily life are either gravitational or electromagnetic. Rounding out the list of fundamental forces are the strong and weak nuclear forces.

If all the forces acting on an object produce no net change, the object is in a state of **equilibrium**. If the object is moving relative to us, we say that it is in a state of **dynamic equilibrium**. If the object is at rest relative to us, we say that it is in a state of **static equilibrium**. Figure 3.1 illustrates these ideas.

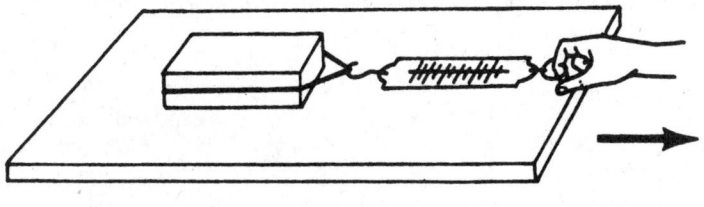

Figure 3.1

Newton's Laws of Motion

In 1687, at the urging of his friend Edmund Halley, Isaac Newton published his greatest work. It was titled the *Mathematical Principles of Natural Philosophy* but is more widely known by a shortened version, *Principia*. In this book, Newton revolutionized the rational study of mechanics by the introduction of mathematical principles that all of nature was considered to obey. Using his newly developed ideas, Newton set out to explain the observations and analyses of Galileo Galilei and Johannes Kepler.

The ability of an object to resist a change in its state of motion is called **inertia**. This concept is the key to Newton's first law of motion:

> *Every body continues in its state of rest, or of uniform motion in a straight line, unless it is compelled to change that state by forces acting on it.*

In other words, an object at rest will tend to stay at rest, and an object in motion will tend to stay in motion, unless acted upon by an external force. By "rest," we of course mean the observed state of rest in a particular frame of reference. As stated above, the concept of "inertia" is taken to mean the ability of an object to resist a force attempting to change its state of motion. As we will subsequently see, this concept is covered under the new concept of **mass** (a scalar, as opposed to **weight**, a vector force).

If a mass has an unbalanced force incident upon it, the velocity of the mass is observed to change. The magnitude of this velocity change depends inversely on the amount of mass. In other words, a force directed along the direction of motion will cause a smaller mass to accelerate more than a larger mass. Newton's second law of motion expresses these observations as follows:

> *The change of motion is proportional to the net applied force, and is in the direction which that force acts.*

The acceleration produced by the force is in the same direction as the force. The second law is sometimes expressed as: $\vec{F} = m\vec{a}$.

However, to preserve the vector nature of the forces, and the fact that by "force" we mean "net force," we write the second law as:

$$\vec{F}_{net} = \sum \vec{F} = m\vec{a}$$

The units of force are **newtons**; 1 newton (N) is defined as the force needed to give a 1-kilogram mass an acceleration of 1 meter per second squared. Thus, 1 newton equals 1 kilogram-meter per second squared.

Newton's third law of motion is crucial for understanding the conservation laws we will discuss later. It stresses the fact that forces are the result of mutual interactions and are thus produced in pairs. Sometimes referred to as the "action-reaction" law, the third law can be stated as follows (Figure 3.2):

TIP
Make sure you understand all of Newton's laws. On an AP free-response problem, you should make a declaration like $\sum \vec{F} = 0$ (for equilibrium) or $\sum \vec{F} = m\vec{a}$ (for accelerated systems).

TIP
The normal force is NOT the reaction force to gravity.

For every force, there is an equal but opposite force on the object causing the original force.

Objects A and B are in contact or colliding.

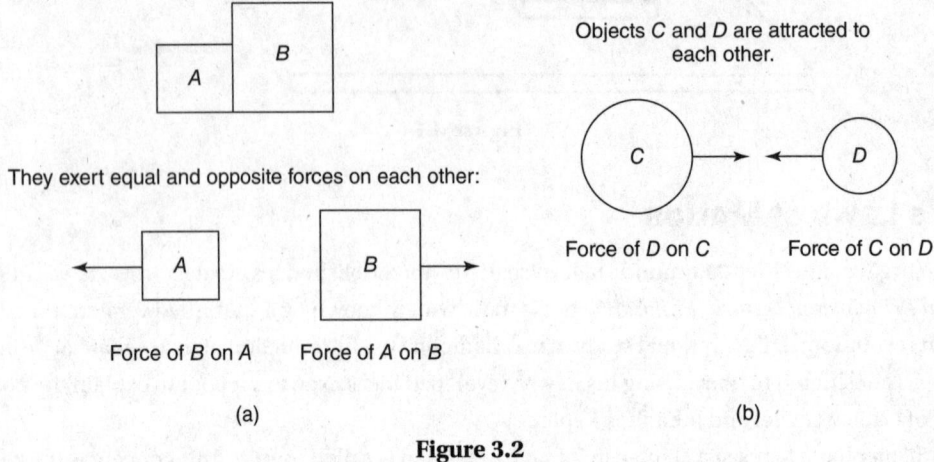

They exert equal and opposite forces on each other:

Force of B on A   Force of A on B

(a)

Objects C and D are attracted to each other.

Force of D on C   Force of C on D

(b)

Figure 3.2

Remember these tips for Newton's third law action-reaction pairs.

- There are no unpaired forces in the universe. Recall that all forces arise from an interaction between two objects.
- Action and reaction forces are applied to two different objects. The pair of forces do not cancel each other for this reason.
- One force does not cause the other. Which force is the action and which is the reaction is a matter of perspective. They happen simultaneously.
- Action-reaction force pairs are the same type of force. In other words, the reaction to a normal force must be another normal force, in the opposite direction, on the other surface. For example, a normal force would never be the reaction force to a gravitational force.

Sample Problem

A 2-kg mass is accelerated by a net force of 20 N. What is the acceleration of the mass?

Solution

We use Newton's second law:

$$\vec{F}_{net} = m\vec{a}$$

Thus,

$$a = \frac{20 \text{ N}}{2 \text{ kg}} = 10 \text{ N/kg} = 10 \text{ m/s}^2$$

Sample Problem

How much does a 2-kg mass weigh on the surface of Earth?

Solution

We use the formula for weight:

$$F_g = mg$$
$$F_g = (2 \text{ kg})(9.8 \text{ m/s}^2) = 19.6 \text{ N}$$

Notice in both problems that the units for acceleration can be written as both N/kg or m/s². They are equivalent. We can think of the acceleration due to gravity in terms of the force per unit mass (which is referred to as a field strength).

Sample Problem

Spencer is discussing orbital motion with Oliver. Oliver claims that Earth must be exerting a stronger gravitational force back on the Moon since the Moon orbits Earth and not the other way around. Spencer points out that this cannot be the case because the two gravitational forces are an action-reaction pair and thus must be equal in magnitude. Oliver maintains that cannot be true since Earth is much bigger than the Moon.

Which student is making the better argument? Explain your answer.

Solution

Spencer is correct. The gravitational pull of Earth on the Moon and that of the Moon on Earth is an action-reaction pair. These forces are the same interaction, and one gravitational attraction could not exist without the other. The different sizes or masses of the two objects do not create different forces for each other—the Moon and Earth each experience the same force. What would change, however, is the acceleration each feels in response to the force. The Moon, with its lower inertia, experiences a much larger change in its motion than Earth.

Static Applications of Newton's Laws

If we look more closely at Newton's second law of motion, we see an interesting implication. If the net force acting on an object is zero, the acceleration of the object will likewise be zero. Notice, however, that kinematically, zero acceleration does not imply zero motion! It simply indicates that the velocity of the object is not changing. If the object is in a state of rest and remains at rest (because of zero net forces), the object is in static equilibrium. Some interesting problems in engineering deal with the static stability of structures. Let's look at a simple example.

> **Think About It**
>
> It is not enough simply to say that two variables are confounded. You must be able to explain the possible incorrect conclusion in context!

Place this book on a table. You will observe that it is not moving relative to you. Its state of rest is provided by the zero net force between the downward force of gravity and the upward reactive force of the table (pushing on the floor, which in turn pushes up on the table, etc.). Figure 3.3 shows this setup. The upward reactive force of the table, sometimes called the **normal force**, always acts perpendicular to the surface.

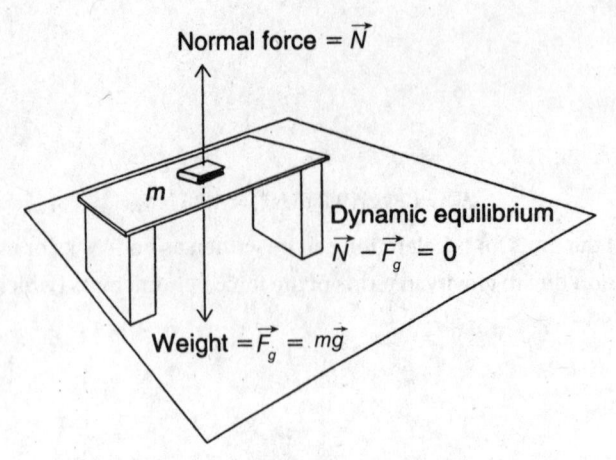

Figure 3.3

Remember: The second law is a vector equation, and so we must treat the sum of all vector forces in each direction separately! In this case, we have:

$$\sum F_y = N - F_g = 0$$

In this example, the normal force is equal to the object's weight. When accelerating upward, the normal force will be greater than the object's weight. Likewise, the normal force will less than the object's weight when accelerating downward. Since the push of the normal force is what the object experiences, the magnitude of the normal force is also called the "apparent weight" of the object.

There are no forces in the *x*-directions to be "analyzed."

A slightly more complex problem would concern a mass suspended by two strings. The tensions in the strings support the mass and are thus vector forces acting at angles. In the problem illustrated in Figure 3.4, a 10-kilogram mass is suspended by two strings, making angles of 30 and 60 degrees, respectively, to the horizontal. The question is, "What are the tensions in the two strings?"

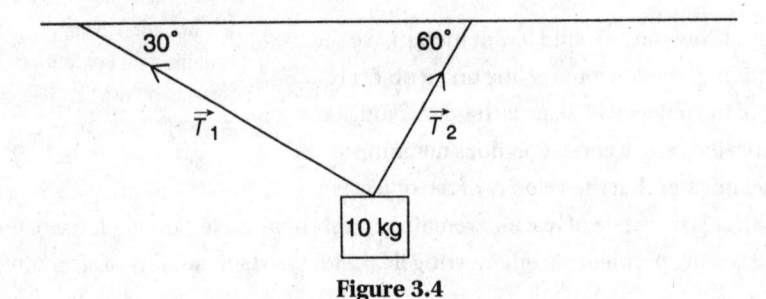

> **TIP**
> The normal force is not always equal to an object's weight. It must be found via Newton's second law.

Figure 3.4

Free-Body Diagrams

A useful heuristic for solving these types of problems is to construct what is called a "free-body diagram." In such a diagram, you "free" the body of its realistic constraints and redraw it, indicating the directions of all applied forces.

You must understand that the College Board has very specific expectations for free-body diagrams. The central object experiencing the forces should be represented by a round dot. All forces experienced by the object should be represented by one arrow coming out from that dot with no stacking (forces in the same direction should be next to each other) and no components indicated. The net force, acceleration, and velocity vector are *not* part of the free-body diagram. For example, free-body diagrams for Figures 3.3 and 3.4 would be

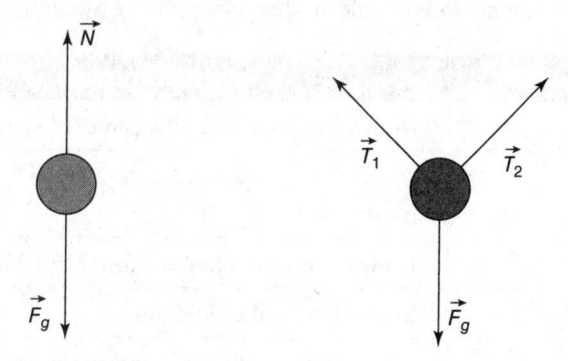

You then use that diagram to set up your static equations for Newton's second law in both the *x*- and *y*-directions. The result is a system of two equations and two unknowns. The two representations of the free-body diagram in Figure 3.5 are useful in identifying the horizontal and vertical components of the forces involved for Figure 3.4. However, these pictures of the components should not be mistaken for an actual free-body diagram.

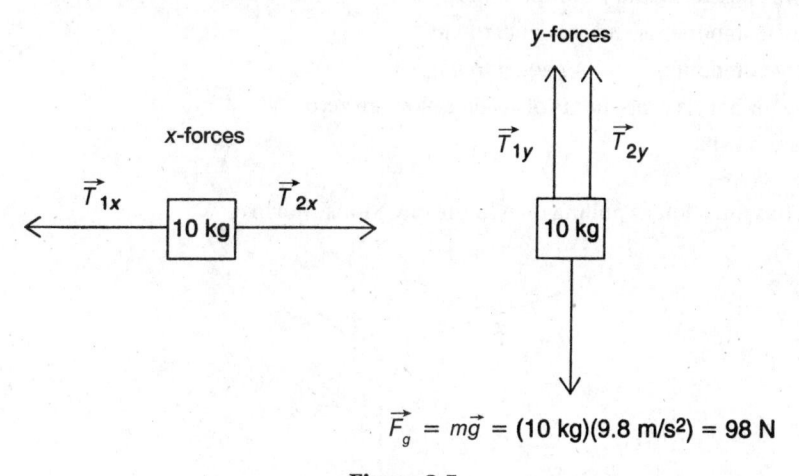

$$\vec{F}_g = m\vec{g} = (10 \text{ kg})(9.8 \text{ m/s}^2) = 98 \text{ N}$$

Figure 3.5

We can now set up our equations for Newton's second law, using the techniques of vector analysis reviewed in Chapter 1. We thus determine the *x*- and *y*-components of the tensions, which are given by $T_{1x} = -T_1 \cos 30$ and $T_{2x} = T_2 \cos 60$. The negative sign is used since T_{1x} is to the left. Thus, in the *x*-direction we write $\sum F_x = ma_x = 0$ since there is no horizontal acceleration.

$$\sum F_x = 0 = T_2 \cos 60 - T_1 \cos 30 = 0.5 T_2 - 0.866 T_1$$

Likewise, for the *y*-direction, we see that the 10-kg mass weighs 98 N. The downward pull of gravity is compensated for by the upward pull of the two vertical components of the tensions. In other words, we can write for the *y*-direction $\sum F_y = ma_y = 0$ since there is no vertical acceleration.

$$\sum F_y = 0 = T_1 \sin 30 + T_2 \sin 60 - F_g = 0.5 T_1 + 0.866 T_2 - 98$$

The two simultaneous equations can then be solved for T_1 and T_2. Performing the algebra, we find that $T_1 = 49$ N and $T_2 = 85$ N.

How to Draw Effective Free-Body Diagrams

The forces, their standard abbreviations, and the rule for their direction are as follows:

Force	Abbreviation	Rule for Direction
Weight	F_g	Drawn straight toward the center of Earth
Normal	F_N	Drawn straight away from the surface (perpendicular)
Tension	F_T	Drawn in the direction the rope is attached
Push or pull	F_P	Made clear by the situation
Friction	F_f	Opposes the motion (velocity vector) or the likely motion if stationary

1. Draw a force diagram/free-body diagram
2. Break every force into its x- and y-components
3. Add up all the x-components and set equal to ma_x
4. Add up all the y-components and set equal to ma_y
5. Decide if some (or both) components of acceleration are zero
6. Solve for the unknowns

Example: An object has three forces pulling on it in the directions shown.

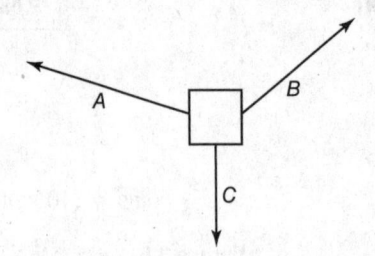

Choose your x-axis and y-axis based on the motion of the object. (One axis should be in the direction of acceleration!) Break vectors A and B into components using trigonometry. Then add up the x-components separately from the y-components.

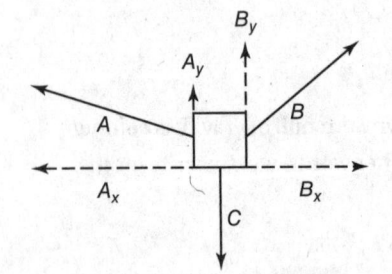

In the chart a sign is added in front of each component to emphasize the importance of dealing with direction.

	A	B	C	$F_{net} = ma$
x-components	$-A_x$	$+B_x$	0	$B_x - A_x = ma_x$
y-components	$+A_y$	$+B_y$	$-C$	$B_y + A_y - C = ma_y$

Use the final column as the starting point for solving the particular problem. At least one of the components of acceleration will be zero in almost all cases. Since we only have two equations, there can only be two unknowns!

Another static situation occurs when a mass is hung from an elastic spring. It is observed that, when the mass is attached vertically to a spring that has a certain natural length, the amount of stretching, or elongation, is directly proportional to the applied weight. This relationship, known as **Hooke's law**, supplies a technique for measuring static forces. Mathematically, Hooke's law is given as

$$\vec{F} = -k\vec{x}$$

where k is the spring constant in units of newtons per meter (N/m), and $\vec{x}$ is the elongation beyond the natural length. The negative is used to indicate that the applied force is restorative so that, if allowed, the spring will accelerate back in the opposite direction. As a static situation, a given spring can be calibrated for known weights or masses, and thus used as a "scale" for indicating weight or other applied forces.

Dynamic Applications of Newton's Laws

If the net force acting on an object is not zero, Newton's second law implies the existence of an acceleration in the direction of the net force. Therefore, if a 10-kilogram mass is acted on by a 10-newton force from rest, the acceleration will be 1 meter per second squared. Heuristically, we can construct a free-body diagram for the system to analyze all forces acting in all directions and then apply the second law of motion. We must be careful, however, to choose an appropriate frame of reference and coordinate system. For example, if the mass is sliding down a frictionless incline, a natural coordinate system to use is one that is rotated in such a way that the x-axis is parallel to the incline (see Figure 3.6).

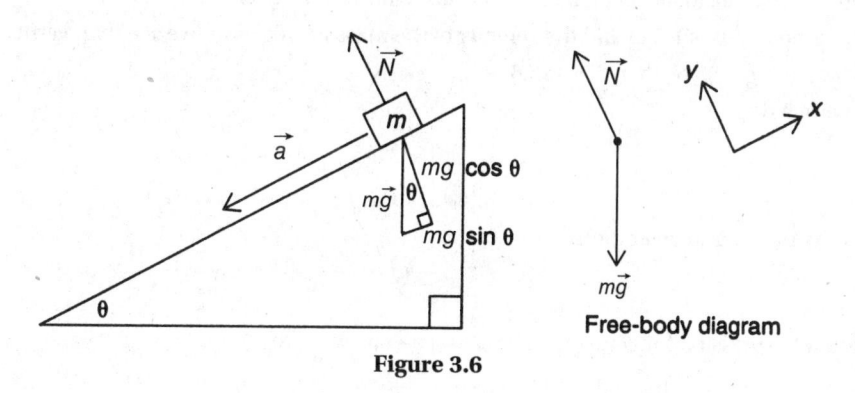

Figure 3.6

Therefore, in order to resolve the force of gravity into components parallel and perpendicular to the incline, we must first identify the relevant angle in the geometry. This is outlined in Figure 3.6, and we can see that the component of weight along the incline is given by $-mg\sin\theta$, where θ is the angle of the incline. The magnitude of the normal force, perpendicular to the incline, is therefore given by $|\vec{N}| = |mg\cos\theta|$ since it is canceling the other component of weight.

If, for example, the mass was 10 kg, the weight would be 98 N. If $\theta = 30°$, we would write for the x-forces:

$$\sum F_x = ma = -mg\sin\theta = -98\sin 30 = -49 \text{ N}$$

giving us an acceleration of 4.01 m/s² down the incline. Since there is no acceleration in the y-direction, we would have:

$$\sum F_y = 0 = N - mg\cos\theta = N - 98\cos 30 = N - 84.9$$

Thus, the normal force is equal to 85 N.

Central Forces

Consider a point mass moving around a circle, supported by a string making a 45-degree angle to a vertical post (the so-called **conical pendulum**). Let's analyze this situation, which is shown in Figure 3.7.

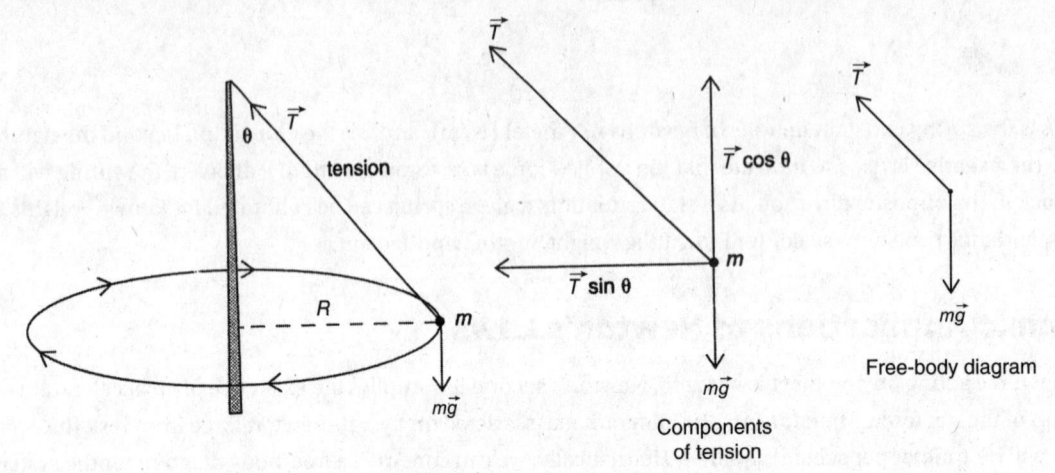

Figure 3.7

In this case, the magnitude of the weight, mg, is balanced by the upward component of the tension in the string, given by $T\cos\theta$. The inward component of tension, $T\sin\theta$, is responsible for providing a **centripetal force**. Note that the centripetal force is *not* an additional force. It is just a description of existing force(s).

If we specify that the radius is 1.5 m and the velocity of the mass is unknown, we see that, vertically, $T\cos 45 = mg$. This implies that, if $mg = 98$ N, then $T = 138.6$ N.

Horizontally, we see that

$$\Sigma F_x = ma_x$$
$$T\sin 45 = ma_c$$

where a_c denotes a centripetal acceleration.

$$a_c = \frac{v^2}{r}$$

Therefore, the general expression for centripetal force is given by

$$F_c = \frac{mv^2}{r}$$

using the logic of $\vec{F} = m\vec{a}$. Using our known information, we solve for velocity and find that $v = 3.8$ m/s. It is important to remember that the components of forces must be resolved along the principal axes of the chosen coordinate system.

> **Sample Problem**

An object of mass m is swung in a vertical circle at a constant speed, s, at the end of a string of length L. Find an expression for the difference in the tensions between the top and the bottom of the circle.

Solution

The tensions will not be equal at the top and the bottom of the circle since the force of gravity is playing a different role at each point:

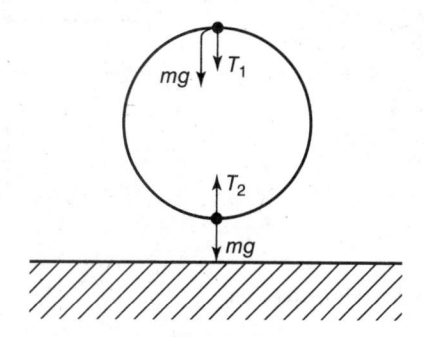

Both T_1 and mg are in the centripetal direction at the top, resulting in the expression:

$$F_{net} = T_1 + mg = ma_c = ms^2/L$$

On the bottom, however, gravity is working against the tension (in the centrifugal direction):

$$F_{net} = T_2 - mg = ma_c = ms^2/L$$

Solving T_1 and T_2, respectively, and taking the difference:

$$T_1 - T_2 = (ms^2/L - mg) - (ms^2/L + mg) = -2\,mg$$

The tension at the top is less by exactly twice the weight of the object.

Friction

Friction is a contact force between two surfaces that is responsible for opposing sliding motion. Even the smoothest surfaces are microscopically rough, with peaks and valleys like a mountain range. If a spring balance is attached to a mass and then pulled, the reading of the force scale before the mass first begins to move provides a measure of the static friction. Once the mass is moving, if we maintain a steady enough force, the velocity of movement will be constant. Thus, the acceleration will be zero, indicating that the net force is zero. The reading of the scale will then measure the kinetic friction. This reading will generally be less than the starting reading.

Observations show that the frictional force is directly related to the applied load pushing the mass into the surface. From our knowledge of forces, the normal force is measuring this "push" into the surface. The proportionality constant linking the normal force with the frictional force is called the coefficient of friction and is symbolized by μ. There are two coefficients of friction: one for static friction (μ_s) and the other for kinetic friction (μ_k). This linear relationship is often written as

$$f_k = \mu_k N$$

and the analysis of a situation involves identifying the normal force.

For example, in Figure 3.8 a mass is being pulled along a horizontal surface by a string, making an angle θ.

> **TIP**
> The coefficient of static friction is generally larger than the coefficient of kinetic friction. Static friction can be any value up to $\mu_s |\vec{N}|$ ($f_s \leq \mu_s |\vec{N}|$).

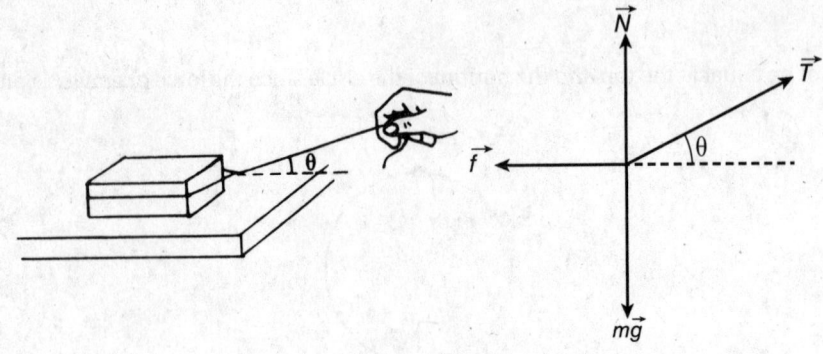

Figure 3.8

Newton's second law in the vertical direction gives us:

$$\sum F_y = m a_y = 0 \quad \text{(no vertical acceleration)}$$
$$N + T\sin\theta - mg = 0$$

Solving for N:

$$N = mg - T\sin\theta$$

Note that the normal force is less than the weight of the mass by precisely the upward component of the tension. Newton's second law in the horizontal direction gives us:

$$\sum F_x = m a_x$$
$$T\cos\theta - f = m a_x$$

Using our model for friction and substituting our expression for the normal force found above:

$$f = \mu_k N = \mu_k (mg - T\sin\theta)$$

With this expression for friction, we can substitute into our horizontal force equation to find the general solution:

$$T\cos\theta - \mu_k(mg - T\sin\theta) = m a_x$$

These types of problems will usually either give all the parameters (T, m, θ, μ_k) and ask for the resulting acceleration or will state the mass is being pulled with constant velocity ($a_x = 0$), allowing you to solve for a different unknown (typically μ_k).

CHAPTER 3: FORCES AND NEWTON'S LAWS OF MOTION

> **Sample Problem**

A 2-kg mass is held at rest at the top of an incline that makes a 40° angle with the ground. The incline is 1.5 m long and the coefficient of kinetic friction $\mu_k = 0.2$. When the mass is released, it slides down the incline with uniform acceleration.

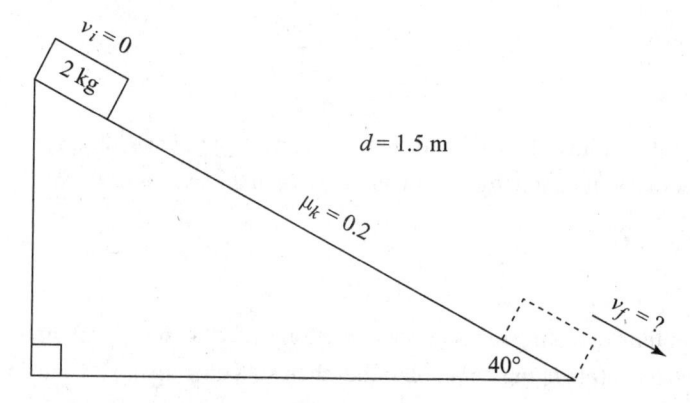

(a) Calculate the acceleration of the mass.
(b) Calculate the speed of the mass when it reaches the bottom of the incline.

> **Solution**

(a)

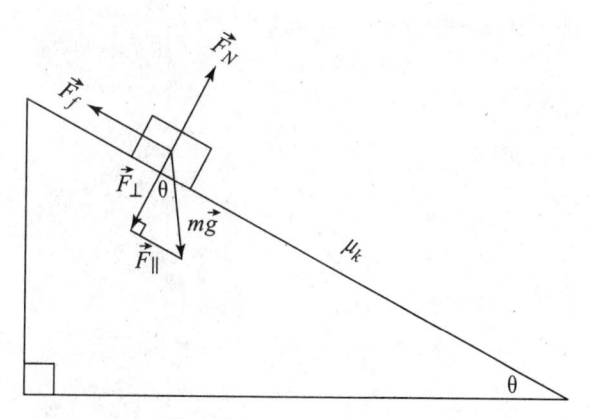

From the geometry, we see that $F_\parallel = mgF\sin\theta$ and $F_\perp = mg\cos\theta$.

We also know that $F_f = \mu_k F_N$.

Taking y to be perpendicular to the plane and x to be down along the plane:

$\sum F_y = ma_y$
$0 = 0$ (since $a_y = 0$)
$N = mg\cos\theta$
$\sum F_x = ma_x$
$mg\sin\theta - F_f = ma_x$
$mg\sin\theta - \mu_k N = ma_x$
$mg\sin\theta - \mu_k mg\cos\theta = ma_x$ (substituting our expression for N above)
$a_x = g(\sin\theta - \mu_k \cos\theta)$ (canceling mass and factoring out g)
$a_x = 4.8 \text{ m/s}^2$ (substituting all our given values)

(b) $v_f^2 = v_i^2 + 2ad$

Since the object starts from rest, $v_i = 0$.
$$v_f^2 = 2ad = 2(4.8 \text{ m/s}^2)(1.5 \text{ m})$$
$$v_f = 3.8 \text{ m/s}$$

> Sample Problem

A car (mass M) takes a turn (of radius R) on a flat road without any slipping or sliding at a constant speed S. Find an expression for the minimum coefficient of friction needed between the wheels and road.

✓ Solution

A rolling tire that is not slipping or sliding relies on static friction to accelerate the car. In this case, the required acceleration is a centripetal one (changing direction rather than speeding up or slowing down):

$$F_{net} = F_f = Ma_c = mS^2/R$$

We can determine an expression for the minimum coefficient of static friction required to take this turn:

$$F_f = mS^2/R$$
$$\mu_s F_N = mS^2/R$$

Since the normal force is only responsible for canceling the weight of the car in this case:

$$\mu_s mg = mS^2/R$$
$$\mu_s = S^2/(gR)$$

Be careful about the changing roles of the forces in vertical circular motion! Radially inward forces are positive and outward ones are negative. Compare the expressions for net force (centripetal force) at the top and bottom of a roller coaster loop.

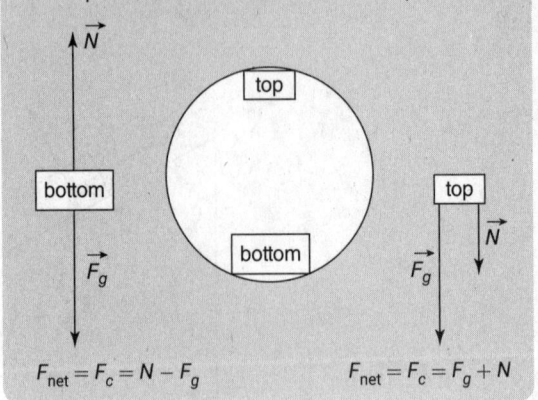

$$F_{net} = F_c = N - F_g \quad\quad F_{net} = F_c = F_g + N$$

SUMMARY

- Forces are pushes and pulls that can be represented by vectors.
- Inertia is the tendency of a mass to resist a force changing its state of motion in a given frame of reference.
- The normal force is a force acting perpendicular to, and away from, a surface.
- Friction is a force that opposes relative motion between two objects.
- Newton's three laws help us to understand dynamics, the actions of forces on masses:
    i. In the absence of an external net force, an object maintains constant velocity (law of inertia).
    ii. If a net force is acting on a mass, then the acceleration is directly proportional to the net force ($\vec{F}_{net} = m\vec{a}$).
    iii. For every action force, there is an equal but opposite reaction force.
- Weight is a force caused by the pull of Earth's gravity on a mass and is directed downward.
- An inertial frame of reference is a frame of reference moving at a constant velocity.
- Free-body diagrams assist in the analysis of forces by identifying and labeling all forces acting on a mass freed from the confines of the illustrated situation.
- In uniform circular motion, the net force is directed inward toward the center and is referred to as the centripetal force. A centrifugal force is a fictitious or pseudo outward force identified in an accelerated frame of reference attached to a mass undergoing circular motion.

Problem-Solving Strategies for Forces and Newton's Laws of Motion

The key to solving force problems, whether static or dynamic, is to construct the proper free-body diagram. Remember that Newton's laws operate on vectors, so you must be sure to resolve all forces into components once a frame of reference and a coordinate system have been chosen. Therefore, you should:

1. Choose a coordinate system. Direct one axis in the direction of acceleration, if known.
2. Make a sketch of the situation if one is not provided.
3. Construct a free-body diagram for the situation.
4. Resolve all forces into perpendicular components based on the chosen coordinate system.
5. Write Newton's second law as the sum of all forces in a given direction. If the problem involves a static situation, set the summation equal to zero. If the situation is dynamic, set the summation equal to ma. Include only applied forces in the diagram.
6. Find the normal force. It is always perpendicular to the surface.
7. Remember that the centripetal force is always directed inward toward the circular path and parallel to the plane of the circle. Gravity is always directed vertically downward.
8. Carefully solve your algebraic equations, using the techniques for simultaneous equations.

Practice Exercises

Multiple-Choice

1. In the situation shown below, what is the tension in string 1?

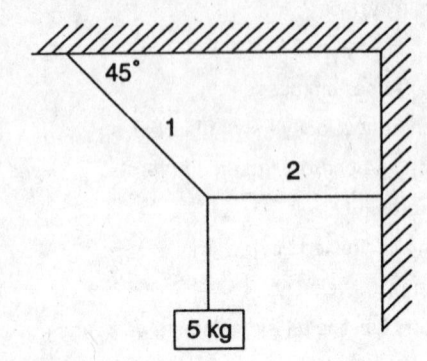

   (A) 69.3 N
   (B) 98 N
   (C) 138.6 N
   (D) 147.6 N

2. Two masses, M and m, are hung over a massless, frictionless pulley as shown below. If $M > m$, what is the downward acceleration of mass M?

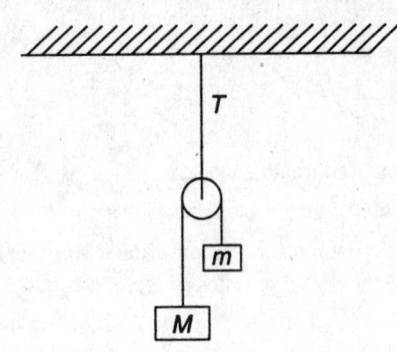

   (A) g
   (B) $\dfrac{(M-m)g}{M+m}$
   (C) $\left(\dfrac{M}{m}\right)g$
   (D) $\dfrac{Mmg}{M+m}$

3. A 0.25-kg mass is attached to a string and swung in a vertical circle whose radius is 0.75 m. At the bottom of the circle, the mass is observed to have a speed of 10 m/s. What is the magnitude of the tension in the string at that point?

   (A) 2.45 N
   (B) 5.78 N
   (C) 22.6 N
   (D) 35.7 N

4. A car and driver have a combined mass of 1,000 kg. The car passes over the top of a hill that has a radius of curvature equal to 10 m. The speed of the car at that instant is 5 m/s. What is the force of the hill on the car as it passes over the top?

   (A) 7,300 N up
   (B) 7,300 N down
   (C) 12,300 N up
   (D) 12,300 N down

5. A hockey puck with a mass of 0.3 kg is sliding along ice that can be considered frictionless. The puck's velocity is 20 m/s. The puck now crosses over onto a floor that has a coefficient of kinetic friction equal to 0.35. How far will the puck travel across the floor before it stops?

   (A) 3 m
   (B) 87 m
   (C) 48 m
   (D) 58 m

6. A spring with a stiffness constant $k = 50$ N/m has a natural length of 0.45 m. It is attached to the top of an incline that makes a 30° angle with the horizontal. The incline is 2.4 m long. A mass of 2 kg is attached to the spring, causing it to be stretched down the incline. How far down the incline does the end of the spring rest?

   (A) 0.196 m
   (B) 0.45 m
   (C) 0.646 m
   (D) 0.835 m

7. A 20-N force is pushing two blocks horizontally along a frictionless floor as shown below.

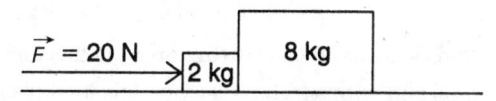

   $\vec{F} = 20$ N → [2 kg] [8 kg]

   What is the force that the 8-kg mass exerts on the 2-kg mass?

   (A) 4 N
   (B) 8 N
   (C) 16 N
   (D) 20 N

8. A force of 20 N acts horizontally on a mass of 10 kg being pushed along a frictionless incline that makes a 30° angle with the horizontal, as shown below.

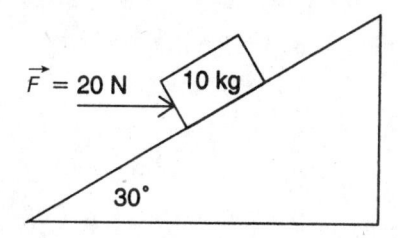

   $\vec{F} = 20$ N → 10 kg, 30°

   The magnitude of the acceleration of the mass up the incline is equal to

   (A) 1.9 m/s²
   (B) 2.2 m/s²
   (C) 3.17 m/s²
   (D) 3.87 m/s²

9. According to the diagram below, what is the tension in the connecting string if the table is frictionless?

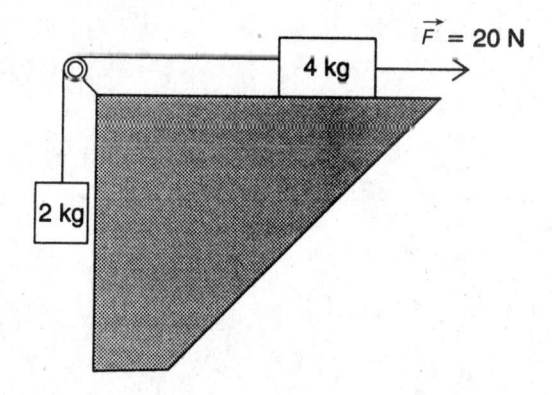

   4 kg, $\vec{F} = 20$ N, 2 kg

   (A) 6.4 N
   (B) 13 N
   (C) 19.7 N
   (D) 25 N

10. A mass, M, is released from rest on an incline that makes a 42° angle with the horizontal. In 3 s, the mass is observed to have gone a distance of 3 m. What is the coefficient of kinetic friction between the mass and the surface of the incline?

    (A) 0.8
    (B) 0.7
    (C) 0.6
    (D) 0.5

Additional Practice

1. A 3-kg block is placed on top of a 7-kg block as shown below.

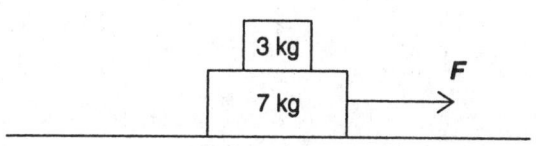

   [3 kg on top of 7 kg], F →

   The coefficient of kinetic friction between the 7-kg block and the surface is 0.35. A horizontal force F acts on the 7-kg block.

   (a) Draw a free-body diagram for each block.
   (b) Calculate the magnitude of the applied force F necessary to maintain an acceleration of 5 m/s².
   (c) Find the minimum coefficient of static friction necessary to prevent the 3-kg block from slipping.

2. A curved road is banked at an angle θ, such that friction is not necessary for a car to stay on the road. A 2,500-kg car is traveling at a speed of 25 m/s, and the road has a radius of curvature equal to 40 m.

   (a) Draw a free-body diagram for the situation described above.
   (b) Find angle θ.
   (c) Calculate the magnitude of the force that the road exerts on the car.

3. The "rotor" is an amusement park ride that can be modeled as a rotating cylinder, with radius r. A person inside the rotor is held motionless against the sides of the ride as it rotates with a certain velocity. The coefficient of static friction between a person and the sides is μ.

   (a) Derive a formula for the period of rotation T, in terms of r, g, and μ.
   (b) If $r = 5$ m and $\mu = 0.5$, calculate the value of the period T in seconds.
   (c) Using the answer to part (b), calculate the angular velocity ω in radians per second.

4. How does the rotation of Earth affect the apparent weight of a 1-kg mass at the equator?

5. If forces occur in action-reaction pairs that are equal and opposite, how is it possible for any one force to cause an object to move?

Answers Explained

Multiple-Choice Problems

1. **(A)** We need to apply the second law for static equilibrium. This means that we must resolve the tensions into their x- and y-components. For T_1, we have $T_1 \cos 45$ and $T_1 \sin 45$. In equilibrium, the sum of all the x forces must equal zero. This means that $T_1 \cos 45 = T_2$ (since T_2 is entirely horizontal). The y-component of T_1 must balance the weight $= mg = 49$ N. Thus, $T_1 \sin 45 = 49$ N and $T_1 = 69.3$ N.

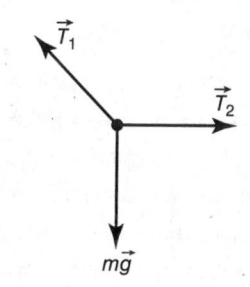

2. **(B)** The free-body diagrams for both masses, M and m, look like this:

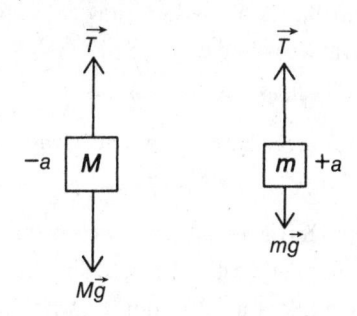

    The large mass is accelerating downward, while the small mass is accelerating upward. The tension in the string is directed upward, while gravity, given by the weight, is directed downward. Using the second law for accelerated motion, we must show that for each mass, $\sum F_y = ma$. Thus, we have

    $$T - Mg = -Ma \quad \text{and} \quad T - mg = ma$$

    Eliminating the tension T and solving for a gives us $(M - m)g/(M + m)$.

3. **(D)** A sketch of the situation is shown below:

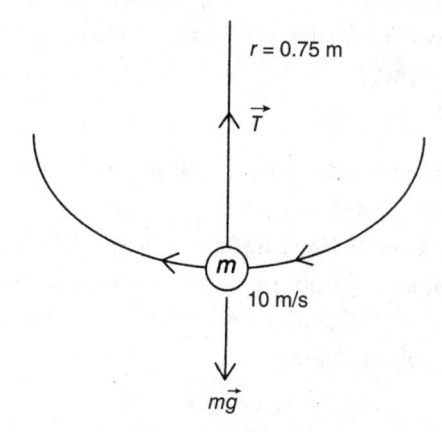

    At the lowest point, the downward force of gravity is opposed by the upward tension in the string. Note that doing a quick tip-to-tail graphical solution of the force vectors must result in an upward (inward) net force. Thus, we can write

    $$T - mg = \frac{mv^2}{r}$$

    Substituting in the given values and solving for T gives us 35.7 N.

4. **(A)** As the car goes over the hill, the force that the hill exerts on the car is the normal force, $\vec{N}$. This is opposed, however, by the downward force of gravity. The combination produces (as in question 3) the centripetal force $\vec{F}_c$. Using our circular motion convention of inward forces being positive, we can write

    $$mg - N = \frac{mv^2}{r}$$

    A sketch is shown below. Note that doing a quick tip-to-tail graphical solution of the force vectors must result in a downward (inward) net force. Substituting in the given values and solving for N gives 7,300 N upward.

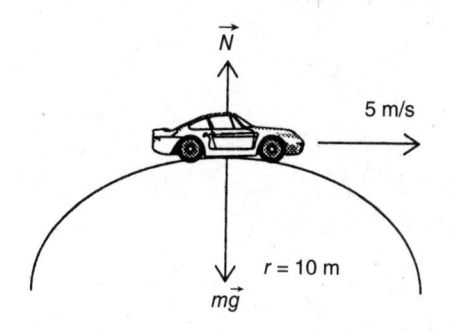

5. **(D)** With horizontal motion, the normal force is equal to the force of gravity. When the puck is moving with constant velocity, no net force is acting on it. Thus, when friction acts to slow the puck down, the friction is the only net force. Therefore, we write

$$f = \mu N = \mu mg = -ma$$

(since the puck is decelerating). Substituting the known numbers gives a deceleration of -3.43 m/s². Now, the final velocity will be zero; and since the time to stop is unknown, we use the following kinematic expression to solve for the stopping distance:

$$-v_i^2 = 2ax$$

Substituting the known numbers, we obtain approximately 58 m as the distance the puck will travel before stopping.

6. **(C)** A sketch of the situation is given below. The spring constant is $k = 50$ N/m.

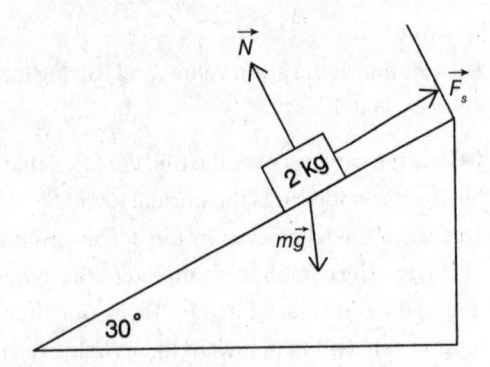

According to Hooke's law, $F_s = kx$, where x is the elongation in excess of the spring's natural length (here 0.45 m). The force, in this case, is provided by the component of weight parallel to the incline, which is given by $mg \sin \theta$. Substituting the known numbers gives us an elongation of $x = 0.196$ m. Since this must be in excess of the spring's natural length, the answer is 0.646 m.

7. **(C)** The 20-N force is pushing on a total mass of 10 kg. Thus, using $\vec{F} = m\vec{a}$, we have the acceleration of both blocks equal to 2 m/s². We draw a free-body diagram for the 2-kg mass as shown below.

Let P represent the force that the 8-kg mass exerts on the 2-kg mass. Writing the second law of motion, we get $\vec{F} - P = m\vec{a}$. To find P, we substitute the known numbers: $20 - P = (2)(2) = 4$. Thus, $P = 16$ N.

8. **(C)** From the given diagram, we see that the force necessary to move the mass up the incline must be in excess of the component force of gravity trying to pull the mass down the incline. The component of gravity down the incline is always given by $mg \sin \theta$. Resolving the given force into a component parallel to the incline and a component perpendicular to the incline, we find that the force up the incline is, at the same angle, $F \cos \theta$. Thus, in general we would write:

$$F \cos \theta - mg \sin \theta = ma$$

Substituting the known numbers gives $a = 3.17$ m/s².

9. **(C)** From the given diagram, we see that the 4-kg mass is accelerating to the right and the 2-kg mass is accelerating up. Thus, we write that the sum of all forces, in each direction, equals ma. In the x-direction, we have (since the tension in the string will try to pull left) $-T + 20 = 4a$. In the y-direction, the tension pulls up against gravity. We therefore write $T - 19.6 = 2a$. Solving for T by eliminating a, we get $T = 19.7$ N.

10. **(A)** We know that the mass accelerates from rest uniformly in 3 s and goes a distance of 3 m. Thus, we can say that

$$d = \frac{1}{2}at^2$$

Substituting the known number gives us an acceleration of 0.67 m/s². On an incline, the normal force is given by $mg\cos\theta$, and so friction can be expressed as $f = \mu mg\cos\theta$. Once again, this force opposes the downward force of gravity parallel to the incline, given by $-mg\sin\theta$. In the downward direction, these two forces are added together and set equal to $-ma$. Thus, we write for this case

$$\mu(M)(9.8)\cos 42 - (M)(9.8)\sin 42 = -(M)(0.67)$$

The masses all cancel out, and solving for μ gives 0.8 as the coefficient of kinetic friction.

Additional Practice

1. (a) A free-body diagram for each of the two blocks is given below:

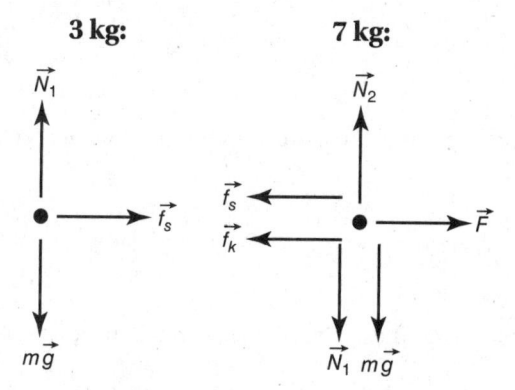

   3 kg: $\vec{N_1}$ up, mg down, $\vec{f_s}$ to the right

   7 kg: $\vec{N_2}$ up, $\vec{N_1}$ and mg down, $\vec{f_s}$ and $\vec{f_k}$ to the left, $\vec{F}$ to the right

   Note the two action-reaction pairs here:
   - $\vec{N_1}$ is up on the 3-kg mass, and $\vec{N_1}$ is down on the 7-kg mass.
   - $\vec{f_s}$ is to the right on the 3-kg mass, and $\vec{f_s}$ is to the left on the 7-kg mass.

   (b) The force of static friction between the two blocks is the force responsible for accelerating the 3-kg block to the right (hence the direction of friction in the free-body diagram). On the other hand, kinetic friction opposes the applied force, F, acting on the 7-kg block. Even so, the force F must accelerate both blocks combined. Thus, in the horizontal direction we can write that, for the second law of motion,

   $$F - f_k = Ma$$

   Also, since $f_k = \mu N$, we can write (since we have horizontal motion, the normal force is equal to the combined weights) $F - (0.35)(10)(9.8) = (10)(5)$. Solving for F, we get $F = 84.3$ N.

   (c) Now, for the two blocks together, we know that static friction provides the force needed for the 3-kg block to accelerate at 5 m/s². The normal force ($\vec{N_1}$) on this surface is just the weight of the 3-kg block, as seen in the free-body diagram drawn in part (a). Thus, $f_s = ma$; $\mu N = ma$; $\mu(3)(9.8) = (3)(5)$. From this we get $\mu = 0.51$.

2. (a) A sketch of the situation and the free-body diagram are given below:

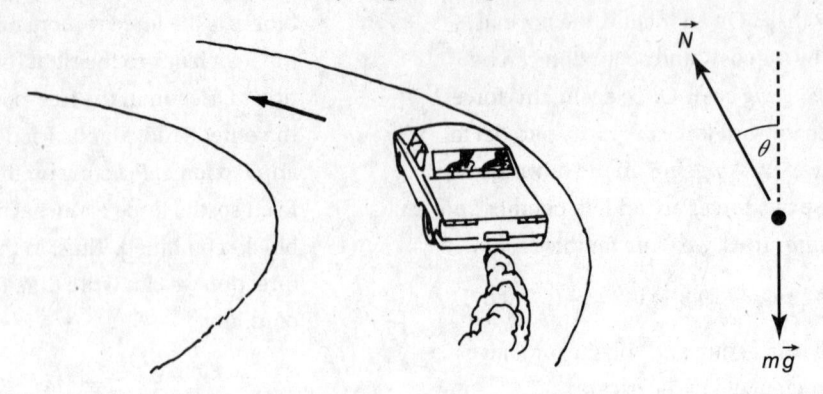

Note that a quick tip-to-tail vector addition of our two forces results in a net force toward the center of the circle of travel: $\vec{F}_N + \vec{F}_g = \vec{F}_{net}$

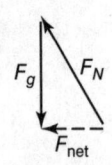

(b) In the coordinates chosen, the horizontal component of the normal force provides the centripetal acceleration. Thus, we can write, in the absence of friction,

$$F_c = ma_c$$

$$N\sin\theta = \frac{mv^2}{r}$$

We can also see from the vertical components that $N\cos\theta = mg$. Thus, eliminating N from both equations gives us

$$\tan\theta = \frac{v^2}{rg}$$

Substituting the known numbers, we find that $\tan\theta = 1.59$ and $\theta = 57.8°$ (rather steep).

(c) Substituting our known value for the angle into the force equation above:

$$N\sin(57.8) = \frac{mv^2}{r}$$

$$N(0.846) = \frac{2,500\,(25)^2}{40}$$

$$N = 46,000 \text{ N}$$

3. (a) A diagram of the rotor is seen below. Here, the normal force is perpendicular to the person and supplies the inward centripetal force. Friction acts along the wall against gravity, which tends to slide the person down. Thus, the key to stability is to be fast enough to maintain equilibrium.

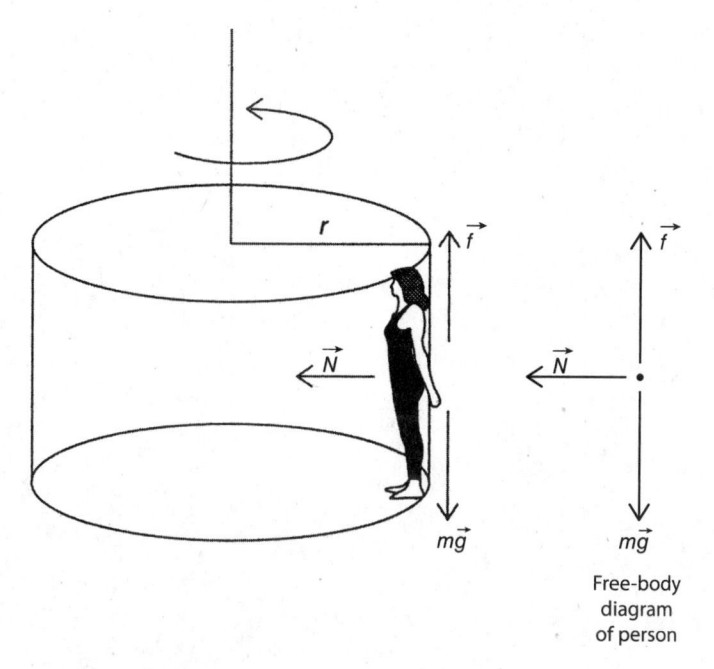

Free-body diagram of person

Substituting in our relationship between velocity and period for circular motion $\left(v = \frac{2\pi r}{T}\right)$ into our expression for centripetal forces, we obtain:

$$F_c = ma_c$$
$$F_c = \frac{mv^2}{r}$$
$$N = \frac{4M\pi^2 r}{T^2}$$

In the vertical direction, $f = Mg$; thus

$$f = \mu N = \frac{\mu M 4\pi^2 r}{T^2} = Mg$$

Solving for T, we get

$$T = \sqrt{\frac{\mu 4\pi^2 r}{g}}$$

This makes sense, since if the coefficient of friction is high, the rotation rate can be small, and thus a larger period!

(b) Substituting the known numbers gives $T = 13.7$ s.

(c) $\omega = \frac{2\pi}{T} = \frac{6.28}{3.17} = 1.98$ rad/s.

4. In the frame of reference of the mass, there is an apparent upward force that tends to reduce the apparent weight of the mass at the equator. This effect is very small, and only very sensitive scales can measure it.

5. This is a classic question and is very tricky. The answer is that the action and reaction pairs act on different objects. Thus, the applied force, if it is (or results in) a net force, can cause an object to move in a given frame of reference. The reaction force, however, is experienced by a different object, which may or may not have its own acceleration due to other forces it is experiencing.

4

Energy

Learning Objectives

In this chapter, you will learn about:
- → Work
- → Power
- → Kinetic energy and the work-energy theorem
- → Potential energy and conservative forces
- → Conservation of energy and systems

Work

When a force is applied to an object over a displacement, **work** is done on the object. The simplest type of work problem involves a single mass m and a constant force F that causes a displacement d. The angle is between the force and displacement vectors. Work is measured in joules and is a scalar; 1 joule = 1 newton · meter:

$$W = Fd \cos\theta$$

❯ Sample Problem 1

In each scenario below, a force F is applied to a box of mass m that is pushed a distance d horizontally across a frictionless table. If the box is displaced by 20 m, calculate the work W done by the 10 N force indicated.

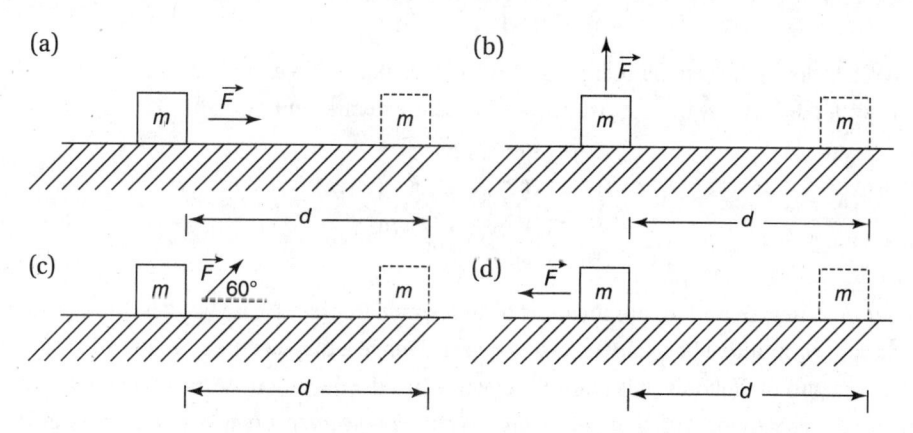

(a) Force $\vec{F}$ applied horizontally in direction of motion.

(b) Force $\vec{F}$ applied vertically upward.

(c) Force $\vec{F}$ applied at 60° above horizontal in direction of motion.

(d) Force $\vec{F}$ applied horizontally opposite to direction of motion.

✓ Solutions

(a) The force and displacement are in the same direction, so all of the force is effective. We use equation $W = Fd$.

$$W = Fd$$
$$W = (10 \text{ N})(20 \text{ m})$$
$$W = 200 \text{ J}$$

> **TIP**
> When calculating work, the mass m of the object being displaced is irrelevant.

Note that $W = Fd$ is a simplified version of $W = (F\cos\theta)d$. The angle is determined by placing both arrows at the origin and measuring the smallest angle between them. If the force and displacement are in the same direction, then $\theta = 0°$. Thus:

$$W = (F\cos\theta)d$$
$$W = (10 \text{ N})(\cos 0°)(20 \text{ m})$$
$$W = (10 \text{ N})(1)(20 \text{ m})$$
$$W = 200 \text{ J}$$

We get the same answer.

(b) The force is not in the same direction as the displacement, so not all of the force is effective. (In fact, none of the force is effective.) We use equation $W = (F\cos\theta)d$. The force is pushing at right angles to the displacement, so $\theta = 90°$:

$$W = (F\cos\theta)d$$
$$W = (10 \text{ N})(\cos 90°)(20 \text{ m})$$
$$W = (10 \text{ N})(0)(20 \text{ m})$$
$$W = 0 \text{ J}$$

(c) Again, the force is not in the same direction as the displacement, so not all of the force is effective. We use equation $W = (F\cos\theta)d$. The force is applied at a 60° angle from the direction of displacement, so $\theta = 60°$:

$$W = (F\cos\theta)d$$
$$W = (10 \text{ N})(\cos 60°)(20 \text{ m})$$
$$W = (10 \text{ N})(0.5)(20 \text{ m})$$
$$W = 100 \text{ J}$$

(d) Again, the force is not in the same direction as the displacement, so we use the equation $W = (F\cos\theta)d$. The force is applied in entirely the opposite direction as the displacement, so $\theta = 180°$:

$$W = (F\cos\theta)d$$
$$W = (10 \text{ N})(\cos 180°)(20 \text{ m})$$
$$W = (10 \text{ N})(-1)(20 \text{ m})$$
$$W = -200 \text{ J}$$

In Sample Problem 1, cases (b) and (d) are the most troublesome. In case (b), no work is done by that particular force. This seems counterintuitive because the box does experience a displacement. Remember that work is not simply "effort" on the part of the force. It is effort that results in a displacement of the object *in the direction of the force*. If the object's displacement is unchanged *in the direction of the force*, no energy was transferred to the object *from that particular force*. The source of the force may be expending energy. Think of yourself holding up a heavy weight at a constant height above the floor. If there is no vertical movement, that energy is not being transferred to the object. Therefore, no work is being done on the object.

In case (d), what does it mean to have negative work? The negative sign tells us that the flow of energy is reversed. Energy is being transferred *from* the mass *to* the source of the force. For example, frictional forces often do negative work, meaning that the mass is losing energy.

Sample Problem 2

Calculate the **total work** (or **net work**) done on object M by the four forces indicated.

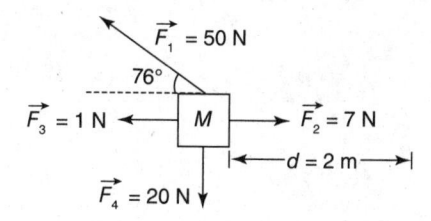

$\vec{F}_1 = 50$ N at $76°$
$\vec{F}_3 = 1$ N
$\vec{F}_2 = 7$ N
$\vec{F}_4 = 20$ N
$d = 2$ m

Solution

Total, or net, work is the sum of the individual works done by each of the individual forces.

$$W_{total} = W_1 + W_2 + W_3 + W_4$$
$$W = (F \cos \theta) d$$
$$W_1 = (50 \text{ N})(\cos 104°)(2 \text{ m})$$
$$W_1 = (10 \text{ N})(-0.24)(2 \text{ m})$$
$$W_1 = -24 \text{ J}$$
$$W_2 = (7 \text{ N})(2 \text{ m})$$
$$W_2 = 14 \text{ J}$$
$$W_3 = (1 \text{ N})(\cos 180°)(2 \text{ m})$$
$$W_3 = (1 \text{ N})(-1)(2)$$
$$W_3 = -2 \text{ J}$$
$$W_4 = (20 \text{ N})(\cos 90°)(2 \text{ m})$$
$$W_4 = (20 \text{ N})(0)(2 \text{ m})$$
$$W_4 = 0$$
$$W_{total} = (-24 \text{ J}) + (14 \text{ J}) + (-2 \text{ J}) + (0 \text{ J})$$
$$W_{total} = -12 \text{ J}$$

REMEMBER

θ is measured relative to the direction of displacement.

In Sample Problem 2, can we compute the net force and then do one work calculation? Try it, and see if you get the same answer. The work done by the net force is equal to the work done by all the individual forces *when you assume that all forces are acting on the center of mass*. Note that this assumption applies in translational motion only but not in rotational motion or deformations of the object.

Power

When work is done over some amount of time, we can measure the **power**, or the rate at which the work is performed. Power is a scalar with units of watts; 1 watt = 1 joule/sec:

$$P = \frac{W}{t}$$

$$P = \frac{\Delta E}{t}$$

> **TIP**
>
> When a constant force *F* is applied to an object moving at constant velocity *v*, the power *P* is the product of the two values: *P = Fv*.

It is an important skill to be able to rewrite any defined unit in terms of the fundamental units of meters, kilograms, seconds, and coulombs. Dimensional analysis of the units of work and power:

Work: Joules = Nm = (kgm/s^2)m = kgm^2/s^2

Power: Watts = J/s = Nm/s = (kgm/s^2)m/s = kgm^2/s^3

Note that the next to last step in the power unit analysis clearly shows units of force times velocity as in the "Tip" on this page.

Sample Problem 3

If the 2-meter displacement in Sample Problem 2 occurred in 4 seconds, how much power was delivered to the object?

Solution

We use the equation $P = \frac{W}{t}$.

$$P = \frac{W}{t}$$

$$P = \frac{-12 \text{ J}}{4 \text{ s}}$$

$P = -3\text{W}$ (The negative sign indicates that the object is losing energy rather than gaining it.)

Kinetic Energy and the Work-Energy Theorem

If work is a measure of energy transferred, what does a gain, or loss, of energy mean? One way energy can be easily seen is through an object's motion. This energy of motion is called **kinetic energy**, *KE*:

$$KE = \tfrac{1}{2}mv^2$$

> **Sample Problem 4**

A 16-kg object starts at rest and gains 2 J of kinetic energy. Determine the final speed of the object.

✓ **Solution**

Since the object started at rest, the change in kinetic energy is exactly the kinetic energy of the moving object:

$$KE = \frac{1}{2}mv^2$$
$$2\,J = \frac{1}{2}(16\text{ kg})v^2$$
$$\frac{1}{4} = v^2$$
$$0.5\text{ m/s} = v$$

If we are calculating the net work done by all external forces, the gain or loss of energy W is seen as a change in the speed of the object, which is incorporated in the object's kinetic energy KE. This is the **work-energy theorem**:

$$W_{total} = \Delta KE$$

> **Sample Problem 5**

A 5-kg object is sliding across a floor at 10 m/s. How much work is done by friction to bring it to a stop?

✓ **Solution**

The change in kinetic energy is the only calculation we can make:

$$KE_i = \text{initial kinetic energy}$$
$$KE_f = \text{final kinetic energy}$$
$$v_i = \text{initial velocity}$$
$$v_f = \text{final velocity}$$

We are told that $v_i = 10$ m/s. Since the object comes to a stop, we know that $v_f = 0$ m/s:

$$W_{total} = \Delta KE = KE_f - KE_i$$
$$W_{total} = \frac{1}{2}mv_f^2 - \frac{1}{2}mv_i^2$$
$$W_{total} = \frac{1}{2}(16\text{ kg})(0\text{ m/s})^2 - \frac{1}{2}(16\text{ kg})(10\text{ m/s})^2$$
$$W_{total} = 0 - 250\text{ J}$$
$$W_{total} = -250\text{ J}$$

The object has lost 250 joules of movement energy as it slowed down. This is the work done by friction: -250 J. Note that if we knew the distance, we could calculate the force required to bring the object to a stop.

Potential Energy and Conservative Forces

Potential energy, *PE* (*PE* is also represented by *U* and is the abbreviation most likely to be seen on the actual exam), refers to energy stored within a system (between particles that are bound by certain types of forces). The forces that are related to potential energy are called **conservative forces**. There are many forms of potential energy. For the purposes of the AP exam, we are concerned with only three: gravitational, elastic (springs), and electrical. Treat all other forces as nonconservative. Generally speaking, we are concerned only with *changes* in potential energies.

Gravity: The farther two masses are separated, the greater the potential energy stored in the system. The mass that is interacting with Earth close to the planet's surface may change the relationship by changing its height:

$$PE_{gravity} = U_g = mgh$$

Elastics (e.g., springs): The more a spring is compressed or stretched away from its equilibrium point, the greater the potential energy stored in the system. The stretched or compressed spring has stored potential energy:

$$PE_{elastic} = \tfrac{1}{2}kx^2$$
$$PE_{elastic} = U_s = \tfrac{1}{2}kx^2$$

For examples, see Chapter 8.

Conservation of Energy and Systems

If only conservative forces are present, the **mechanical energy** (kinetic energy *KE* plus potential energy *PE*) is conserved. The **law of conservation of energy** states that the initial mechanical energy must equal the final mechanical energy.

Initial Mechanical Energy = Final Mechanical Energy
$$KE_i + PE_i = KE_f + PE_f$$

In most cases, the potential energy of mechanical systems that appear on the exam will be limited to gravitational and elastic, so the general form would be:

$$\tfrac{1}{2}mv_i^2 + mgh_i + \tfrac{1}{2}kx_i^2 = \tfrac{1}{2}mv_f^2 + mgh_f + \tfrac{1}{2}kx_f^2$$

▶ Sample Problem 6

A rock falls from a height of 20 meters in a vacuum. How fast is the rock traveling when it reaches the bottom?

Solution

In this problem, we are dealing with the potential energy due to gravity. So $PE = mgh$. Since there are no nonconservative forces in this problem, we use the law of conservation of energy: $KE_i + PE_i = KE_f + PE_f$

From the information given, we know that:

$h_i = 20$ m
$v_i = 0$ m/s
$h_f = 0$ m

$$KE_i + PE_i = KE_f + PE_f$$
$$\tfrac{1}{2}mv_i^2 + mgh_i = \tfrac{1}{2}mv_f^2 + mgh_f$$
$$\tfrac{1}{2}(0 \text{ m/s})^2 + (10 \text{ m/s})^2 (20 \text{ m}) = \tfrac{1}{2}v_f^2 + (10 \text{ m/s})^2 (0 \text{ m})$$
$$200 \text{ m}^2/\text{s}^2 = \tfrac{1}{2}v_f^2$$
$$400 \text{ m}^2/\text{s}^2 = v_f^2$$
$$20 \text{ m/s} = v_f$$

Note that the direction of v_f is downward. Also note that the mass m canceled out completely at the start of the problem-solving.

Sample Problem 7

An arrow is shot from the roof of a building 30 meters high at 5 m/s and at an angle of 45 degrees. How fast will the arrow be going when it hits the ground?

Solution

From the information given, we know that:

$h_i = 30$ m
$v_i = 5$ m/s
$h_f = 0$ m

$$KE_i + PE_i = KE_f + PE_f$$
$$\tfrac{1}{2}mv_i^2 + mgh_i = \tfrac{1}{2}mv_f^2 + mgh_f$$
$$\tfrac{1}{2}v_i^2 + gh_i = \tfrac{1}{2}v_f^2 + gh_f$$
$$\tfrac{1}{2}v_i^2 + gh_i = \tfrac{1}{2}v_f^2 + g(0 \text{ m})$$
$$\tfrac{1}{2}v_i^2 + gh_i = \tfrac{1}{2}v_f^2$$
$$v_f^2 = v_i^2 + 2gh_i$$
$$v_f^2 = (5 \text{ m/s})^2 + (2)(10 \text{ m/s})^2 (30 \text{ m})$$
$$v_f^2 = 25 \text{ m}^2/\text{s}^2 + 600 \text{ m}^2/\text{s}^2$$
$$v_f^2 = 625 \text{ m}^2/\text{s}^2$$
$$v_f = 25 \text{ m/s}$$

Sample Problem 8

A 2-kg mass sliding along a frictionless floor at 3 m/s hits a spring (with $k = 200$ N/m), compresses it, and bounces back. What is the maximum compression of the spring?

✓ Solution

In this problem, we are dealing with the potential energy of a spring. So $PE = \frac{1}{2}kx^2$.

Note that the final state refers to the spring at maximum compression. From the information given, we know:

$m = 2$ kg
$x_i = 0$ m
$v_i = 3$ m/s
$v_f = 0$ m/s

$$KE_i + PE_i = KE_f + PE_f$$
$$\frac{1}{2}mv_i^2 + \frac{1}{2}kx_i^2 = \frac{1}{2}mv_f^2 + \frac{1}{2}kx_f^2$$
$$\frac{1}{2}mv_i^2 + \frac{1}{2}k(0 \text{ m})^2 = \frac{1}{2}m(0 \text{ m/s})^2 + \frac{1}{2}kx_f^2$$
$$\frac{1}{2}mv_i^2 = \frac{1}{2}kx_f^2$$
$$mv_i^2 = kx_f^2$$
$$(2 \text{ kg})(3 \text{ m/s})^2 = (200 \text{ N/m})x_f^2$$
$$x_f^2 = \frac{9}{100} \text{ m}$$
$$x_f = \frac{3}{10} = 0.3 \text{ m}$$

REMEMBER

Velocity is 0 m/s when the spring is at maximum compression ($v = 0$ during any turnaround).

If **nonconservative forces** are present (e.g., friction), the work done by those forces must be accounted for when calculating the final kinetic and potential energies. If W_{NC} is the work done by nonconservative forces, the energies must be balanced in the following manner:

$$KE_i + PE_i + W_{NC} = KE_f + PE_f$$

TIP

On the AP Physics 1 exam, any losses in mechanical energy due to nonconservative forces are assumed to be lost to sound or thermal energy.

Sample Problem 9

A 2-kg rock falls from a cliff 20 m above the ground. The air friction does −39 J of work during the fall. What is the final speed of the rock?

✓ Solution

Since nonconservative forces are involved, we use equation

$$KE_i + PE_i + W_{NC} = KE_f + PE_f$$

From the information given, we know:

$m = 2$ kg
$h_i = 20$ m
$v_i = 0$ m/s
$h_f = 0$ m

$$KE_i + PE_i + W_{NC} = KE_f + PE_f$$

$$\tfrac{1}{2}mv_i^2 + mgh_i + (-39 \text{ J}) = \tfrac{1}{2}mv_f^2 = mgh_f$$

$$\tfrac{1}{2}(2 \text{ kg})(0 \text{ m/s})^2 + (2 \text{ kg})(10 \text{ m/s}^2)(20 \text{ m}) - (39 \text{ J}) = \tfrac{1}{2}(2 \text{ kg})v_f^2 + (2 \text{ kg})(10 \text{ m/s}^2)(0)$$

$$0 + 400 - 39 = v_f^2 + 0$$

$$361 = v_f^2$$

$$19 \text{ m/s} = v_f$$

▶ Sample Problem 10

Julia is attempting to calculate the final speed of a dropped ball on Earth using energy consideration. She starts with the work-energy theorem:

$$W = \Delta E$$

She then considers the ball to have dropped over a vertical distance of Δy starting from rest. So Julia first determines the work done by gravity:

$$W = fd = mg\Delta y$$

She then computes the change in energy (the final energy being kinetic and the initial energy being gravitational potential):

$$\Delta E = E_f - E_i = \tfrac{1}{2}mv^2 - mg\Delta y$$

Julia sets the two equations equal to each other:

$$W = \Delta E$$

$$mg\Delta y = \tfrac{1}{2}mv^2 - mg\Delta y$$

She solves for v:

$$v = \sqrt{4g\Delta y}$$

From simple kinematics, Julia knows this answer is wrong. Find her error, and explain what she should do to correct it.

✓ Solution

Julia has included gravity twice in her work. First she determined the work done by gravity. Then she included gravitational potential energy in her energy term. She must commit to using either

$$W_{net} = \Delta KE$$

(calculate the work done by gravity but do not use gravitational potential energy)

or
$$W_{NC} = \Delta(KE + PE)$$

(use a zero for the work done by nonconservative forces in this example and then include gravitational potential energy)

Once Julia eliminates an extra usage of gravity, the correct answer in either case will agree with kinematics:
$$v = \sqrt{2g\Delta y}$$

Sample Problem 11

A 15-kg bucket is raised from the bottom of a well in 2.5 seconds. The well is 8 meters deep, and the bucket is not in motion at the beginning or end of the lift. What was the power exerted while raising the bucket?

Solution

We use the equation $P = \frac{\Delta E}{t}$, where the energy is gravitational potential energy since this is the only change in the bucket's energy:

$$\Delta E = mg\Delta h = (15)(9.8)(8) = 1{,}176 \text{ J}$$
$$P = \frac{\Delta E}{t} = 1{,}176 \text{ J}/(2.5 \text{ s}) = 470 \text{ J/s} = 470 \text{ W}$$

SUMMARY

- Work (measured in joules) is both a force over a distance and a transfer of energy.
- Power (measured in watts) is the rate at which work is done.
- Net work done by all forces to an object is equal to the change in kinetic energy of that object.
- Work done by conservative forces is path independent.
- Potential energy is stored energy within a system due to internal conservative forces. If you take into account only the work done by nonconservative forces to a system, the work-energy theorem becomes:

$$W_{NC} = \Delta KE + \Delta PE$$

- If there are only conservative forces, then the total mechanical energy ($KE + PE$) is conserved.
- If mechanical energy is not conserved in an isolated system, the "missing" energy can be found in the form of internal energies (thermal energy, etc.).

Problem-Solving Strategies for Energy Problems

Given a system, is the external force that is doing work putting joules of energy into kinetic, potential, or internal energy? Generally speaking, the work could be going into any of these. Here are some guidelines:

1. If all forces doing work on the object are treated as external forces (i.e., no potential energies are being used), the net work is going into kinetic energy only ($W_{net} = \Delta KE$).
2. If the only external forces are gravity and/or springs, those forces can be described as potential energies and the mechanical energy is conserved.
3. If the problem has both conservative and nonconservative forces, the problem itself must offer clues as to where the nonconservative forces are putting the energy: $W_{NC} = \Delta KE + \Delta PE$.
4. Often students will assume that mechanical energy is strictly conserved in a problem (i.e., all the joules of energy are to be found in KE and PE). They will overlook a nongravitational, nonspring force (usually friction) in the problem that takes some energy out of KE and PE and puts it into IE (internal energy).

> Often a problem can be solved either through kinematics or energy. Usually energy is the quicker, more powerful tool. However, time permitting, you can double-check your answer with simple kinematics as long as the acceleration is constant.

Practice Exercises

Multiple-Choice

1. Which of the following are the units for the spring constant, k?

    (A) $kg \cdot m^2/s^2$
    (B) $kg \cdot s^2$
    (C) $kg \cdot m/s$
    (D) kg/s^2

2. Which of the following is an expression for mechanical power?

    (A) Ft/m
    (B) F^2m/a
    (C) Fm^2/t
    (D) F^2t/m

3. A pendulum consisting of a mass m attached to a light string of length ℓ is displaced from its rest position, making an angle θ with the vertical. It is then released and allowed to swing freely. Which of the following expressions represents the velocity of the mass when it reaches its lowest position?

    (A) $\sqrt{2g\ell(1 - \cos\theta)}$
    (B) $\sqrt{2g\ell(\tan\theta)}$
    (C) $\sqrt{2g\ell(\cos\theta)}$
    (D) $\sqrt{2g\ell(1 - \sin\theta)}$

4. An engine maintains constant power on a conveyor belt machine. If the belt's velocity is doubled, the magnitude of its average acceleration

    (A) is doubled
    (B) is quartered
    (C) is halved
    (D) is quadrupled

5. A mass m is moving horizontally along a nearly frictionless floor with velocity v. The mass now encounters a part of the floor that has a coefficient of kinetic friction given by μ. The total distance traveled by the mass before it is slowed by friction to a stop is given by

    (A) $2v^2/\mu g$
    (B) $v^2/2\mu g$
    (C) $2\mu g v^2$
    (D) $\mu v^2/2g$

6. Two unequal masses are dropped simultaneously from the same height. The two masses will experience the same change in

    (A) acceleration
    (B) kinetic energy
    (C) potential energy
    (D) velocity

7. A pendulum that consists of a 2-kg mass swings to a maximum vertical displacement of 17 cm above its rest position. At its lowest point, the kinetic energy of the mass is equal to

    (A) 0.33 J
    (B) 3.33 J
    (C) 33.3 J
    (D) 333 J

8. A 0.3-kg mass rests on top of a spring that has been compressed by 0.04 m. Neglect any frictional effects, and consider the spring to be massless. Then, if the spring has a constant k equal to 2,000 N/m, to what height will the mass rise when the system is released?

    (A) 1.24 m
    (B) 0.75 m
    (C) 0.54 m
    (D) 1.04 m

9. A box is pulled along a smooth floor by a force F, making an angle θ with the horizontal. As θ increases, the amount of work done to pull the box the same distance, d,

   (A) increases
   (B) increases and then decreases
   (C) remains the same
   (D) decreases

10. As the time needed to run up a flight of stairs decreases, the amount of work done against gravity

    (A) increases
    (B) decreases
    (C) remains the same
    (D) increases and then decreases

Additional Practice

1. A 0.75-kg sphere is dropped through a tall column of liquid. When the sphere has fallen a distance of 2.0 m, it is observed to have a velocity of 5 m/s.

   (a) How much work was done by the frictional "viscosity" of the liquid?
   (b) What is the average force of friction during the displacement of 2.0 m?

2. A 15-kg mass is attached to a massless spring by a light string that passes over a frictionless pulley as shown below. The spring has a force constant of 500 N/m and is unstretched when the mass is released. What is the velocity of the mass when it has fallen a distance of 0.3 m?

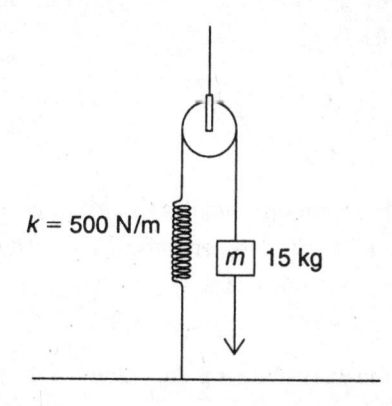

3. A 1.5-kg block is placed on an incline. The mass is connected to a massless spring by means of a light string passed over a frictionless pulley, as shown below. The spring has a force constant k equal to 100 N/m. The block is released from rest, and the spring is initially unstretched. The block moves down a distance of 16 cm before coming to rest. What is the coefficient of kinetic friction between the block and the surface of the incline?

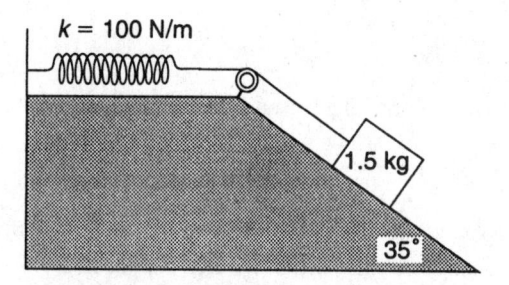

4. Explain how it might be possible for a moving object to possess, and simultaneously not possess, kinetic energy.

5. When you hold up a 10-kg mass with your arms outstretched, you get tired. However, according to physics, you have not done any work! Explain how this is possible.

6. A car (1,200 kg) is traveling up a steep incline (35 degrees with the horizontal) at a speed of 25 mph. How much power is required?

Answers Explained

Multiple-Choice Problems

1. **(D)** The units for the spring or force constant are provided by Hooke's law, $F = kx$, and are newtons per meter. Recall that $1\text{ N} = 1\text{ kg m/s}^2$, so dividing by m gives kg/s^2.

2. **(D)** Power is equal to work done divided by time. Therefore,
$$P = \frac{W}{t} = \frac{Fd}{t}$$
Using some algebra and kinematics, we see that
$$P = \frac{Fd}{t} = Fv = Fat = F\left(\frac{F}{m}\right)t = \frac{F^2 t}{m}$$
You could also get the answer by verifying which expression has units of joules per second.

3. **(A)** Consider the sketch below of the situation:

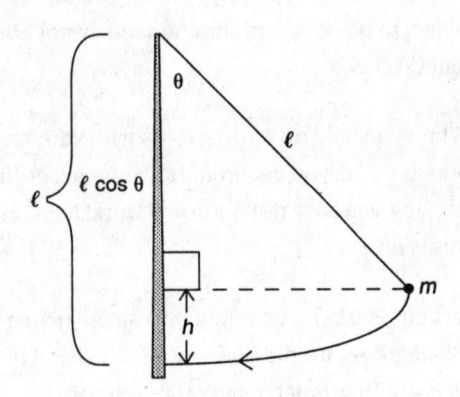

From the geometry of the sketch, note that
$$h = \ell - \ell\cos\theta = \ell(1 - \cos\theta)$$
If it is assumed that there is no friction, gravity is the only conservative force acting to do work. Therefore,
$$ME_{bottom} = ME_{top}$$
$$KE_{bottom} = PE_{top}$$
$$\tfrac{1}{2}mv^2 = mgh$$
and so, at the bottom,
$$v = \sqrt{2gh} = \sqrt{2g\ell(1-\cos\theta)}$$

4. **(C)** Power is equal to the product of the average force applied times the velocity. If the velocity is doubled, and the power is constant, the average force must be halved. Since $F = ma$, the average acceleration of the belt must be halved as well.

5. **(B)** The only applied force is friction, which is doing work to stop the mass. This work is being taken from the initial kinetic energy. For friction, we know that $f = \mu N$, and since the motion is horizontal, $N = mg$. Let x be the distance traveled while stopping; therefore, we can write that $W_f = KE$ and:
$$\tfrac{1}{2}mv^2 = \mu mgx$$
Solving for x, we get:
$$x = \frac{v^2}{2\mu g}$$

6. **(D)** Objects dropped simultaneously from the same height have the same constant acceleration, which is the change in velocity. The unequal masses will provide for different energies, but the velocity changes will be the same.

7. **(B)** Since we have a conservative system,
$$KE_{bottom} = PE_{top}$$
$$= mgh$$
$$= (2)(9.8)(0.17)$$
$$= 3.33\text{ J}$$
Remember to change 17 cm to 0.17 m!

8. **(C)** We are dealing with a conservative system, so the initial starting energy is just the potential energy of the compressed spring. This energy supplies the work needed to raise the mass a height h, which is a gain in gravitational potential energy. Thus, we equate these two expressions and solve for the height:
$$\tfrac{1}{2}(2{,}000)(0.4)^2 = (0.3)(9.8)h$$
Thus, $h = 0.54$ m.

9. **(D)** The component of the applied force in the horizontal direction depends on the cosine of the angle. This value decreases with increasing angle. Thus, the work decreases as well.

10. **(C)** The work done is independent of the time or path taken because gravity is a conservative force. The power generated is affected by time, but the work done to run up the stairs remains the same as long as the same mass is raised to the same height.

Additional Practice

1. (a) The change in potential energy is a measure of the initial energy and equals $(0.75)(9.8)(2) = 14.7$ J. After the sphere has fallen 2 m, its velocity is 5 m/s, so the kinetic energy is given by

   $$KE = \left(\frac{1}{2}\right)(0.75)(5)^2 = 9.375 \text{ J}$$

   The work done by friction is due to a nonconservative force that is equal to the difference between the final and initial energies:

   $$E_f - E_i = 9.375 \text{ J} - 14.7 \text{ J} = -5.325 \text{ J}$$

   Therefore, $W_f = -5.325$ J.

   (b) The average frictional force is equal to the work done divided by the displacement of 2 m. Thus

   $$f = -2.67 \text{ N}$$

   which of course is negative since it opposes the motion.

2. The loss of potential energy is balanced by a gain in elastic potential energy for the spring and in kinetic energy for the falling mass if we assume that the starting energy for the system is zero relative to the starting point for the mass. In the absence of friction, the displacement of the mass is equal to the elongation of the spring. Thus, we can equate our energies and write:

   $$0 = -mgh + \frac{1}{2}kx^2 + \frac{1}{2}mv^2$$

   If we substitute the known numbers, we get

   $$0 = -(15)(9.8)(0.3) + \left(\frac{1}{2}\right)(500)(0.3)^2 + \left(\frac{1}{2}\right)(15)v^2$$

   Solving for velocity v gives $v = 1.7$ m/s.

3. In this problem, the work done by gravity down the incline is affected by the work done by friction. Together, the net work is applied to stretching the spring by an amount equal to the displacement of the mass. Thus, we can say

   $$W_g - W_f = W_s$$

   where W_g is the work done by gravity, W_f is the work done by friction, and W_s is the work done to the spring. Hence:

   $$mg\sin\theta \, d - \mu mg\cos\theta \, d = \frac{1}{2}kx^2$$

   Substituting the known numbers, we get:

   $$(1.5)(9.8)(0.16)\sin 35 - \mu(1.5)(9.8)(0.16)\cos 35 = \frac{1}{2}(100)(0.16)^2$$

   Solving for the coefficient of friction gives $\mu = 0.036$.

4. The motion of an object is always relative to a given frame of reference. If an object is moving relative to one frame, we can envision a second frame, moving with the object, in which it appears to be at rest. Thus, in one frame, the object possesses kinetic energy because of its relative motion. In the second frame, with the object appearing to be at rest, it does not possess kinetic energy.

5. When you hold up a mass, your arm muscles strain against the force of gravity. This requires energy from your body, which makes you feel tired.

6. We will use the power equation when we have a constant force at constant velocity:

   $$P = F_v$$

   First, however, we must convert the speed of 25 mph into m/s:

   25 miles/hour × (1 hour/3,600 seconds) × (1,609 meters/1 mile) = 11 m/s

   Next, we use our knowledge of inclined planes to determine the force the car must be exerting in order to cancel out the down-along-the-incline force of gravity. We know the forces are canceling since the velocity is constant:

   $$F_{car} = mg\sin(35) = 1{,}200(9.8)(0.574) = 6{,}745 \text{ N}$$

   Putting it together:

   $$P = F_v = (6{,}745 \text{ N})(11 \text{ m/s}) = 74{,}000 \text{ W}$$

5

Gravitation

> **Learning Objectives**
>
> In this chapter, you will learn about:
> → Newton's law of universal gravitation
> → Inertial and gravitational mass
> → Gravitational energy
> → Orbiting satellites

Newton's Law of Universal Gravitation

We already know that all objects falling near the surface of Earth have the same acceleration, given by $g = 9.8 \text{ m/s}^2$. This is most easily observed by the independence of the period of a pendulum on the mass of the bob. This empirical verification is independent of any "theory" of gravity.

The weight of an object on Earth is given by $F_g = mg$, which represents the magnitude of the force of gravity due to Earth that is acting on the object. From our discussion in Chapter 2, we know that gravity causes a projectile to assume a parabolic path. Isaac Newton, in his book *Principia*, extended the idea of projectile motion to an imaginary situation in which the velocity of the projectile was so great that the object would fall and fall, but the curvature of Earth would bend away and leave the projectile in "orbit." Newton conjectured that this might be the reason why the Moon orbits Earth. This is shown in Figure 5.1.

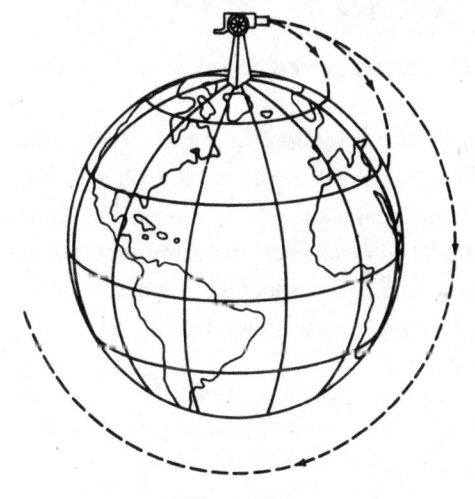

Figure 5.1

To answer this question, Newton first had to determine the centripetal acceleration of the Moon based on observations from astronomy. He knew the relationship between centripetal acceleration and period. We have written that relationship as

$$a_c = \frac{4\pi^2 r}{T^2}$$

where r is the distance to the Moon (in meters) and T is the orbital period (in seconds). From astronomy we know that $r = 3.8 \times 10^8$ m, and since the Moon orbits Earth in 27.3 days, we have $T = 2.3 \times 10^6$ s. Using these values, we find that the magnitude of $a_c = 2.8 \times 10^{-3}$ m/s². Newton realized that some kind of universal gravity was acting as the centripetal force on the Moon.

Using his third law of motion, Newton realized that the force that Earth exerts on the Moon should be exactly equal, but opposite, to the force that the Moon exerts on Earth (observed as tides). This relationship implied that the force should be dependent on the mass of Earth, and this mutual interaction implied that the force of gravity should be proportional to the product of both masses. In other words, Newton's law of gravity could be expressed as

$$F_g = \frac{GM_1 M_2}{r^2}$$

> **TIP**
> Newton's law of universal gravitation is referred to as an inverse square law. For example, if the distance between the masses is doubled, the force between them is one-fourth.

where r is measured from center to center. This is a vector equation in which the force of gravity is directed inward toward the center of mass for the system (in this case, near the center of Earth). Since Earth is many times more massive than the Moon, the Moon orbits Earth, and not vice versa.

The value of G, called the universal gravitational constant, was experimentally determined by Henry Cavendish in 1795. In modern units $G = 6.67 \times 10^{-11}$ N·m²/kg². If we recognize that $F = ma$, and then if we consider M_1 to equal the mass of an object of mass m, and M_2 equal to the mass of Earth, M_E, the acceleration of a mass m in free fall near the surface of Earth is given by

$$F_g = ma$$
$$a = \frac{GM_E}{r_E^2}$$

Substituting in the values for Earth:

$$a = \frac{(6.67 \times 10^{-11})(6.0 \times 10^{24})}{(6.4 \times 10^6)^2} = 9.8 \text{ m/s}^2$$

This is the acceleration of all free-falling objects near the Earth's surface, a value represented simply by the letter "g."

> **TIP**
> Notice that the acceleration due to gravity does not depend on the mass of the falling object. This is consistent with Galileo's observations of falling bodies and the period of a pendulum.

Thus, we have a theory that accounts for the value of the known acceleration due to gravity. In fact, if we replace the mass of Earth by the mass of any other planet, and the radius of Earth by the corresponding radius of the other planet, the formula below allows us to determine the value of g on any planet or astronomical object in the universe! For example, the value of g on the Moon is approximately 1.6 meters per second squared, or about one-sixth the value on Earth. Thus, objects on the Moon weigh one-sixth as much as they do on Earth. This formula also allows us to solve for the acceleration due to gravity at any distance above Earth's surface.

$$g = \frac{GM}{r^2}$$

> **Sample Problem**

(a) Calculate the gravitational force of attraction between a 2,000-kg car and a 12,000-kg truck, separated by 0.5 m.

(b) What happens to this force if the distance between them doubles to 1.0 m?

Solution

(a) We use Newton's law of gravitation:

$$F_g = \frac{GM_1M_2}{r^2}$$

$$F_g = \frac{(6.67 \times 10^{-11} \text{N} \cdot \text{m}^2/\text{kg}^2)(2{,}000 \text{ kg})(12{,}000 \text{ kg})}{(0.5 \text{ m})^2} = 6.4 \times 10^{-3} \text{ N}$$

(b) By doubling the distance, the force will be four times weaker as gravity is an inverse square law. Specifically in this case:

$$0.0064 \text{ N}/4 = 0.0016 \text{ N}$$

Inertial and Gravitational Mass

It is worth noting that mass (m) actually has two roles in physics. Its original role is that of quantifying an object's inertia. For example, we use this inertial mass when calculating the acceleration (a) resulting from a net force (F) on an object: $a = F/m$. However, we also use this same mass when calculating the force of gravity (either in the universal formula shown in this chapter or the simplified $F_{\text{gravity}} = mg$). We do not usually make any distinction between the two roles for they are, in fact, the same mass. This surprising fact that the same quantity is telling us about an object's inertia and about its interaction with gravity is actually a profound clue about how the universe works. Although beyond the scope of the AP Physics 1 exam, this clue is actually one of the things that led Einstein to formulate his general theory of relativity in 1916, which offers an even more complete theory of gravity than Newton's law of universal gravitation.

Gravitational Energy

In Chapter 4, we saw that the amount of work done (by or against gravity) when vertically displacing a mass is given by the change in the gravitational potential energy:

$$\Delta PE = mg\Delta h$$

This formula is good only for h values that are small when compared to r_E and only at the surface of Earth.

We now know that the value of g is not constant but varies inversely with the square of the distance from the center of Earth. Also, since we want the potential energy to become smaller the closer we get to Earth, we can use the results of the previous section to rewrite the potential energy formula as

$$PE = \left(\frac{-GM_0M_E}{r}\right) \text{ (for any } r > r_E)$$

A graph of universal gravitational potential energy (GPE) as a function of distance is shown in Figure 5.2.

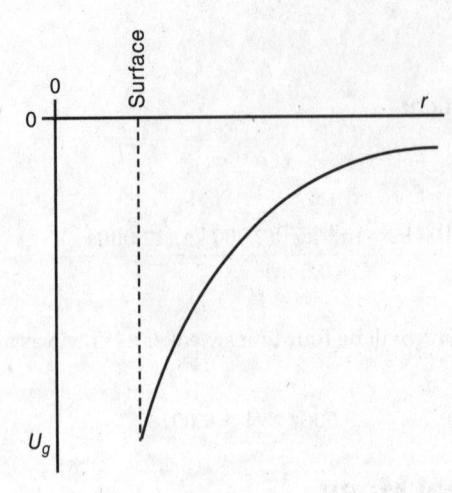

Figure 5.2

As an object gets to higher and higher distances above Earth's surface, the gravitational potential energy does increase like the simplified *mgh*. (It starts out as a large negative number and approaches zero!) The negative values of energy indicate that the object is bound to Earth and can be free only if the object has enough kinetic energy to give it an overall positive mechanical energy.

The "escape velocity" from the gravitational force of Earth can be determined by considering the situation where an object has just reached infinity, with zero final velocity, given some initial velocity at any direction away from the surface of Earth. We designate that escape velocity as v_{esc}, and state that, when the final velocity is zero (at infinity), the total energy must be zero, which implies

$$\frac{1}{2}M_0 v_{esc}^2 = \frac{GM_0 M_E}{r_E}$$

The mass of the object can be eliminated from the relationship, leaving

$$v_{esc} = \sqrt{\frac{2GM_E}{r_E}}$$

An object can leave the surface of Earth with any velocity. There is, however, one minimum velocity at which, if the spacecraft coasted, it would not fall back to Earth because of gravity.

Think About It

Notice that both the **escape velocity** and the **orbital velocity** depend on the mass of Earth and not on the mass of the object.

Orbiting Satellites

The minimal orbital velocity can be determined by assuming that we have an approximately circular orbit (see Chapter 2). In this case, we can set the centripetal force equal to the gravitational force:

$$\frac{GM_0 M_E}{r_E^2} = \frac{M_0 v_0^2}{r_E}$$

Eliminating the mass of the object from the equation leaves

$$v_{orbit} = \sqrt{\frac{GM_E}{r_E}}$$

By comparing the two expressions for escape velocity and orbital velocity, we see that

$$v_{esc} = \sqrt{2}\, v_{orbit}$$

Of course, we could repeat our derivation of orbital speed for any planet of arbitrary mass M and any arbitrary orbital radius r:

$$\boxed{v_{orbit} = \sqrt{\frac{GM}{r}}}$$

By assuming the orbit is circular, we can substitute in $2\pi r/T$ for v_{orbit}. By setting these two expressions for orbital speed equal to each other, we will be able to derive one of Kepler's original laws of planetary motion, which he codified in 1619. Kepler's third law of planetary motion was that all orbiting objects (orbiting around the same central mass) have the same ratio of T^2/r^3. Although Kepler found this relationship to be the same for all orbiters, he could not explain why, nor did he know what the value represented. But, with the laws of Newton, we can show that the ratio is based on the mass of the central object:

$$2\pi r/T = v_{orbit} = \sqrt{\frac{GM}{r}}$$

Squaring both sides and collecting like terms:

$$\frac{T^2}{r^3} = \frac{4\pi^2}{GM} = \text{Constant}$$

On the AP Physics 1 exam, all satellites are assumed to have a mass so much smaller than that of the central mass about which it is orbiting that we can safely assume there is no change in motion of the central object. In this case, a statement for the total energy for the satellite would be

$$E = \tfrac{1}{2}mv^2 - \frac{GMm}{r}$$

where m is the mass of the satellite and M is the mass of the central object. In the case of the perfectly circular orbit described above, both kinetic and potential energies of the satellite are constant. In most actual orbits, satellites do not maintain perfectly circular motion but, rather, orbit in ellipses. To conserve total energy, the kinetic energy will increase as the satellite approaches its closest point (lowering its potential energy). Likewise, the satellite will be going slowest at its farther point in its elliptical orbit (lowest kinetic energy when its potential energy is highest).

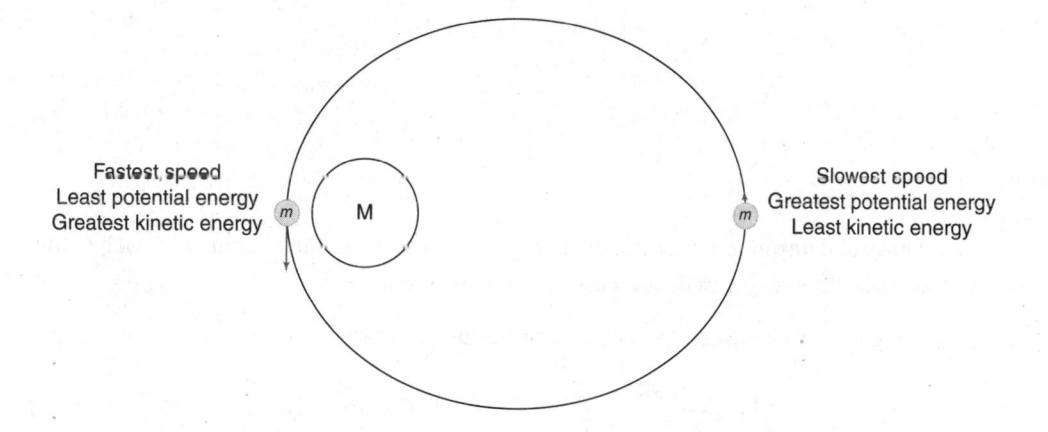

Fastest speed
Least potential energy
Greatest kinetic energy

Slowest speed
Greatest potential energy
Least kinetic energy

> **Sample Problem**

(a) What is the magnitude of the acceleration due to gravity at an altitude of 400 km above the surface of Earth?
(b) What percentage loss in the weight of an object results?

✓ **Solution**

(a) The formula for the acceleration due to gravity above Earth's surface is

$$g = \frac{GM_E}{(r_E + h)^2}$$

We have $M_E = 5.98 \times 10^{24}$ kg, $r_E = 6.38 \times 10^6$ m, $h = 400$ km $= 0.4 \times 10^6$ m. Substituting these values, as well as the known value for G given earlier in this chapter, we get $g = 8.67$ m/s².

(b) The fractional change in weight is found by comparing the value of g at 400 km to its value at Earth's surface: $8.67/9.8 = 0.885$. Thus, there is an 11.5% loss of weight at that height.

> **Sample Problem**

Three uniform spheres of masses **1 kg**, **2 kg**, and **4 kg** are placed at the corners of a right triangle as shown below. The positions relative to the coordinate system indicated are also shown. What is the magnitude of the resultant gravitational force on the 4-kg mass if we consider that mass to be fixed?

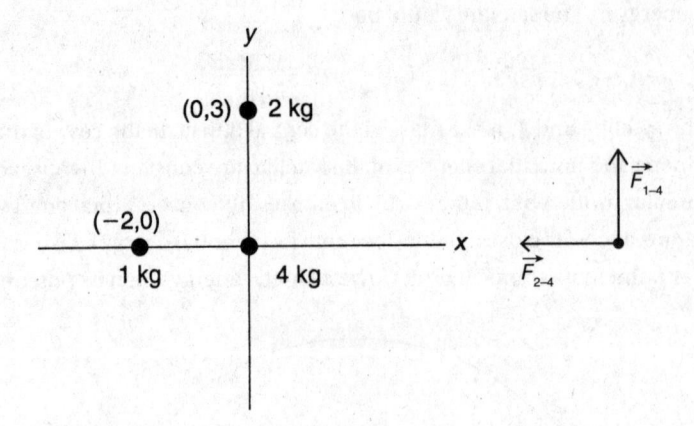

✓ **Solution**

In this problem, we have to determine separately the force of gravitational attraction between each of the smaller masses and the 4-kg mass. Then we must do a vector sum of these forces.

We begin by calculating the force between the 1-kg mass and the 4-kg mass:

$$F_{1-4} = \frac{(6.67 \times 10^{-11})(1)(4)}{2^2} = 6.67 \times 10^{-11} \text{ N}$$

The direction is to the right and is therefore considered positive.

For the 2-kg mass and the 4-kg mass, the downward direction is taken as negative:

$$F_{2-4} = \frac{(6.67 \times 10^{-11})(2)(4)}{3^2} = 5.93 \times 10^{-11} \text{ N}$$

The magnitude of the resultant force is given by the Pythagorean theorem since F_{1-4} is horizontal and F_{2-4} is vertical.

$$|\vec{F}| = \sqrt{(6.67 \times 10^{-11})^2 + (5.93 \times 10^{-11})^2} = 8.92 \times 10^{-11} \text{ N}$$

Sample Problem

Thomas is studying orbital motion. He is trying to determine what new speed a satellite would have if its altitude above the ground was doubled and if its mass was doubled. He starts by using the equation for orbital velocity:

$$v = \sqrt{\frac{GM}{r}}$$

He then reasons that replacing M with $2M$, and r with $2r$, would leave him with the same expression for orbital speed. He concludes that all satellites orbit with the same speed.

Is Thomas correct? If not, what mistakes has he made? Justify your answer.

Solution

Thomas has made two separate mistakes in his analysis. The first is that the M in the orbital velocity equation is for the mass of the central object. The orbiting object's mass has canceled out in the derivation of this formula and is no longer relevant. So the conclusion that the mass of the satellite does not affect orbital speed is correct (but not for the reason he gave). Thomas's second error is that the r in the equation is from the center of the planet to the center of the satellite. So a doubling of altitude (which is measured from Earth's surface) would be less than a doubling of the actual radius of the original orbit. Generally speaking, satellites at higher altitudes orbit at lower speeds. They do this, however, as a function of orbital radius, not altitude.

SUMMARY

- Newton's law of gravitation states that there is a force of attraction between any two masses and the magnitude of this force is directly proportional to the product of the masses and inversely proportional to the square of the distance between them.
- Newton's law of gravitation is sometimes referred to as an inverse square law.

Problem-Solving Strategies for Gravitation

Keep in mind that gravity is a force and therefore a vector quantity. Distances are measured in meters from the center of masses of each object; therefore, if an object is above the surface of Earth, you must add the height to the radius of Earth. If you remember that the force of gravity is an inverse-square-law relationship, you may be able to deduce an answer from logic rather than algebraic calculations, which take time.

Practice Exercises

Multiple-Choice

1. What is the value of g at a height above Earth's surface that is equal to the radius of Earth?

   (A) 9.8 N/kg
   (B) 4.9 N/kg
   (C) 6.93 N/kg
   (D) 2.45 N/kg

2. A planet has half the mass of Earth and half the radius. Compared to the acceleration due to gravity near the surface of Earth, the acceleration of gravity near the surface of this other planet is

   (A) twice as much
   (B) one-fourth as much
   (C) half as much
   (D) the same

3. Which of the following is an expression for the acceleration of gravity at the surface of a planet with uniform density ρ and radius r?

   (A) $G(4\pi\rho/3r^2)$
   (B) $G(4\pi\rho r^2/3)$
   (C) $G(4\pi\rho/3r)$
   (D) $G(4\pi r\rho/3)$

4. What is the orbital velocity of a satellite at a height of 300 km above the surface of Earth? (The mass of Earth is approximately 6×10^{24} kg, and its radius is 6.4×10^6 m.)

   (A) 5.42×10^1 m/s
   (B) 1.15×10^6 m/s
   (C) 7.7×10^3 m/s
   (D) 6×10^6 m/s

5. What is the escape velocity from the Moon, given that the mass of the Moon is 7.2×10^{22} kg and its radius is 1.778×10^6 m?

   (A) 1.64×10^3 m/s
   (B) 2.32×10^3 m/s
   (C) 2.69×10^6 m/s
   (D) 5.38×10^6 m/s

6. A black hole theoretically has an escape velocity that is greater than or equal to the velocity of light (3×10^8 m/s). If the effective mass of the black hole is equal to the mass of the Sun (2×10^{30} kg), what is the effective "radius" (called the "Schwarzschild radius") of the black hole?

   (A) 3×10^3 m
   (B) 1.5×10^3 m
   (C) 8.9×10^6 m
   (D) 4.45×10^6 m

7. What is the gravitational force of attraction between two trucks, each of mass 20,000 kg, separated by a distance of 2 m?

   (A) 0.057 N
   (B) 0.013 N
   (C) 0.0067 N
   (D) 1.20 N

8. The gravitational force between two masses is 36 N. If the distance between masses is tripled, the force of gravity will be

   (A) the same
   (B) 18 N
   (C) 9 N
   (D) 4 N

Additional Practice

1. Find the magnitude of the gravitational field strength g at a point P along the perpendicular bisector between two equal masses, M and M, that are separated by a distance $2b$ as shown below:

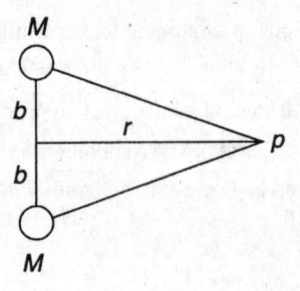

2. Explain why a heavier object near the surface of Earth does not fall faster than a lighter object (neglect air resistance).

3. Show that the units for g in N/kg are equivalent to m/s^2. Why is g referred to as the "gravitational field strength"?

4. Explain why objects in orbit appear to be "weightless."

5. Find the orbital radius for a geosynchronous satellite. A geosynchronous satellite is a satellite around Earth that is synchronized, in that its orbital period is 24 hours, just like Earth itself. The mass of Earth is approximately 6.0×10^{24} kg.

Answers Explained

Multiple-Choice Problems

1. **(D)** The value of g varies inversely with the square of the distance from the center of Earth; therefore, if we double the distance from the center (as in this case), the value of g decreases by one-fourth: 1/4(9.8) = 2.45. Since $g = F/m$, alternative units are newtons per kilogram.

2. **(A)** If, using the formula for g, we take half the mass, the value decreases by one-half. If we decrease the radius by half, the value will increase by four times. Combining both effects results in an overall increase of two times.

3. **(D)** The formula for g is $g = GM/r^2$. The planet is essentially a sphere of mass M, radius r, and density ρ (with $M = V\rho$, where V is the volume). The volume of the planet is given by $V = 4/3\pi r^3$. Making the substitutions yields:

$$g = G\frac{4\pi r\rho}{3}$$

4. **(C)** The formula for orbital velocity is

$$v_{orbit} = \sqrt{\frac{GM}{r}}$$

where r is the distance from the center of Earth. In this case, we must add 300 km = 300,000 m to the radius of Earth. Thus, $r = 6.7 \times 10^6$ m. Substituting the given values yields $v_{orbit} = 7,728$ or 7.7×10^3 m/s.

5. **(B)** The formula for escape velocity is

$$v_{esc} = \sqrt{\frac{2GM_{Moon}}{r_{Moon}}}$$

Substituting the given values yields $v_{esc} = 2,324$ or 2.32×10^3 m/s.

6. **(A)** From question 6, we know that the escape velocity is given by $v_{esc} = \sqrt{2GM/r}$. To find r, we need to square both sides, and then solve for the radius. This yields $r = 2GM/v^2$. Substituting the given values yields $r = 3,000$ or 3×10^3 m. (This is only a "theoretical" size for the black hole. As an interesting exercise, try calculating the value of g on such an object!)

7. **(C)** We use the formula for gravitational force:

$$F = \frac{GM_1 M_2}{r^2}$$

Substituting the given values (don't forget to square the distance!) yields $F = 0.0067$ N.

8. **(D)** The force of gravity is an inverse-square-law relationship. This means that, as the distance is tripled, the force is decreased by one-ninth. One-ninth of 36 N is 4 N.

Additional Practice

1. Let's designate as g_1 the field strength at P caused by the top mass and designate as g_2 the field strength at point P due to the bottom mass. Both of these field strengths are accelerations and therefore vectors. The distance from point P to the line connecting the masses is r, and the midpoint distance connecting the masses is b. Therefore, the distance from point P to each mass is given by the Pythagorean theorem and is equal to $\sqrt{r^2 + b^2}$.

The direction of each acceleration g is directed toward each mass from point P. Since each mass is identical and the distance to each mass is the same, the angles formed by the vectors to the x-axis are the same. Let's call each angle θ such that $\tan\theta = b/r$ in magnitude.

From the above analysis we can conclude that

$$g_1 = g_2 = \frac{GM}{r^2 + b^2}$$

The vector components of each field strength result in a symmetrical cancellation of the y-components. This is true because the direction of g_1 is toward the upper left, and thus its x-component is directed to the left and its y-component is directed upward. Field strength g_2 has an x-component that is also directed to the left and equal in magnitude to the x-component of g_1. The y-component of g_2 is directed downward and is also equal in magnitude to the y-component of g_1. Since these two vectors are equal and opposite, they will sum to zero and will not contribute to the net resultant field (which is just directed horizontally to the left).

What remains to be done is to determine the expression for the x-component of g_1 or g_2 and then multiply by 2. Since $g_{1x} = g_1 \cos\theta$, we can see from the geometry that

$$\cos\theta = \frac{r}{\sqrt{r^2 + b^2}}$$

Combining results gives

$$g_{net} = \frac{2GMr}{(r^2 + b^2)^{3/2}}$$

2. The acceleration due to gravity near the surface of Earth is given by

$$g = \frac{GM_E}{r^2}$$

Hence, this acceleration is independent of the mass of an object.

3. We know that $F = ma$, and for gravity, $F = mg$. Hence,

$$g = \frac{F}{m} = N/kg$$

But we also know that $N = kg \cdot m/s^2$. Thus, in units,

$$N/kg = m/s^2$$

A gravitational field measures the amount of force per unit mass. Thus, g is referred to as the gravitational field strength since $g = F/m$.

4. Objects in orbit are in free fall. Thus, all objects fall together and appear to be weightless. A person in free fall therefore does not experience any contact forces and does not feel his or her own weight.

5. We know the orbital period of this satellite must be 24 hours:

$$T = 24 \text{ hrs} \times (3{,}600 \text{ sec/hour}) = 86{,}400 \text{ s}$$

Using Newton's version of Kepler's third law:

$$\frac{T^2}{r^3} = \frac{4\pi^2}{GM} = \text{Constant}$$

Knowing T for our satellite, the mass of the Earth (6.0×10^{24} kg), and the universal gravitational constant (G), we can solve for r:

$$r = (GMT^2/4\pi^2)^{1/3}$$
$$r = 4.2 \times 10^7 \text{ m}$$

6
Impacts and Linear Momentum

Learning Objectives

In this chapter, you will learn about:
- → Internal and external forces
- → Impact forces and momentum changes
- → The law of conservation of linear momentum
- → Elastic and inelastic collisions
- → Center of mass

Internal and External Forces

Consider a system of two blocks with masses m and M ($M > m$). If the blocks were to collide, the forces of impact would be equal and opposite. However, because of the different masses, the response to these forces (i.e., the changes in velocity) would not be equal. In the absence of any outside or external forces acting on the objects (such as friction or gravity), we say that the impact forces are internal.

To Newton, the "quantity of motion" discussed in his book *Principia* was the product of an object's mass and velocity. This quantity, called **linear momentum** or just **momentum**, is a vector quantity having units of kilogram · meters per second (kg · m/s). Algebraically, momentum is designated by the letter $\vec{p}$, such that $\vec{p} = m\vec{v}$.

> **TIP**
> Momentum and impulse are both vector quantities.

To understand Newton's rationale, consider the action of trying to change the motion of a moving object. Do not confuse this with the inertia of the object; here, both the mass and the velocity are important. Consider, for example, that a truck moving at a slow 1 meter per second can still inflict a large amount of damage because of its mass. Also, a small bullet, having a mass of perhaps 1 gram or less, does incredible damage because of its high velocity. In each case (see Figure 6.1), the damage is the result of a force of impact when the object is intercepted by something else. Let's now consider the nature of impact forces.

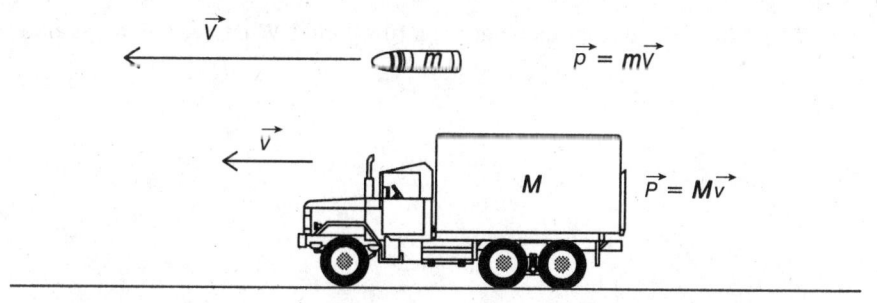

Figure 6.1 Comparison of the Momentum of a Truck with That of a Bullet

Impact Forces and Momentum Changes

Consider a mass m moving with a velocity v in some frame of reference. If the mass is subjected to some external forces, then, by Newton's second law of motion, we can write that $\sum F = ma$. The vector sum of all the forces, referred to as the **net force**, is responsible for changing the velocity of the motion (in magnitude and/or direction).

If we recall the definition of acceleration as the rate of change or velocity, we can rewrite the second law of motion as

$$\vec{F}_{net} = m\left(\frac{\Delta \vec{v}}{\Delta t}\right)$$

This expression is also a vector equation and is equivalent to the second law of motion. If we make the assumption that the mass of the object is not changing, we can again rewrite the second law in the form

$$\vec{F}_{net} = \frac{\Delta m\vec{v}}{\Delta t} = \frac{\Delta \vec{p}}{\Delta t}$$

This expression means that the net external force acting on an object is equal to the rate of change of momentum of the object and is another alternative form of Newton's second law of motion. This **change in momentum** is a vector quantity in the same direction as the net force applied. Although a change in mass can change the momentum, on the AP Physics 1 exam, mass is assumed to be constant. Since the time interval is just a scalar quantity, we can multiply both sides by Δt to get

$$\vec{F}_{net} \Delta t = \Delta \vec{p} = m\Delta \vec{v} = m\vec{v}_f - m\vec{v}_i$$

The quantity, $\vec{F}_{net} \Delta t$, also represented by $\vec{J}$, called the **impulse**, represents the effect of a force acting on a mass during a time interval Δt, and is likewise a vector quantity. From this expression, it can be stated that the impulse applied to an object is equal to the change in momentum of the object.

Another way to consider impulse is to look at a graph of force versus time for a continuously varying force (see Figure 6.2).

The area under this curve is a measure of the impulse in units of newton · seconds (N · s). Another way to view this concept is to identify the average force, $\vec{F}_{avg}$, such that the area of the rectangle formed by the average force is equal in area to the entire curve. This is more manageable algebraically, and we can write

$$\vec{J} = \vec{F}_{avg} \Delta t = \Delta \vec{p}$$

> **TIP**
>
> Although the units are the same, impulse is not simply equal to momentum; impulse is the *change* in momentum, which is similar to how work has the same units as energy but is equal to the *change* in energy!

Sample Problem

A 2,000-kg car is traveling at 20 m/s and stops moving over a 10-s period. What was the magnitude of the average braking force?

Solution

We know that

$$\vec{F} = \frac{m\Delta \vec{v}}{\Delta t}$$

Substituting, we obtain

$$F = \frac{(2{,}000 \text{ kg})(20 \text{ m/s})}{10 \text{ s}} = 4{,}000 \text{ N}$$

Figure 6.2

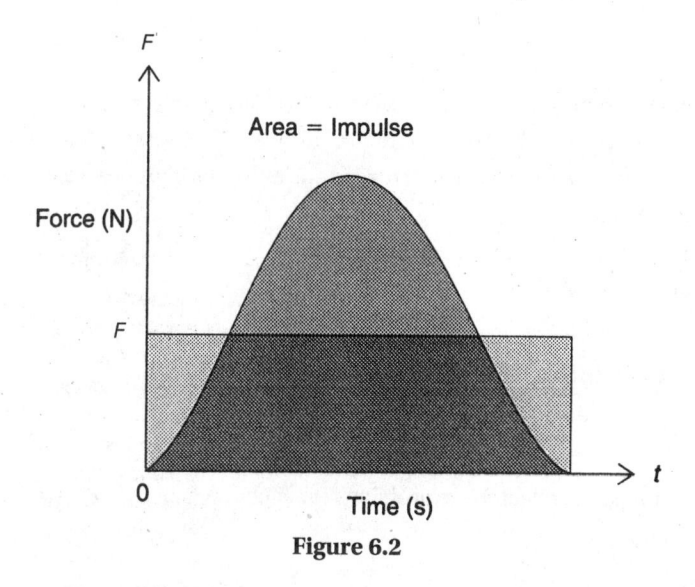

Area = Impulse
Force (N) vs Time (s)

> How do airbags work? After all, the same impulse must be supplied to the car's occupants during the crash. The airbag lengthens the time interval over which the force is applied. The area under the curve is the same. However, it is wider and not as high (lower force).

The Law of Conservation of Linear Momentum

Newton's third law of motion states that for every action there is an equal but opposite reaction. This reaction force is present whenever we have an interaction between two objects in the universe. Suppose we have two masses, m_1 and m_2, that are approaching each other along a horizontal frictionless surface. Let $\vec{F}_{12}$ be the force that m_1 exerts on m_2, and let $\vec{F}_{21}$ be the force that m_2 exerts on m_1. According to Newton's law, these forces must be equal and opposite; that is, $\vec{F}_{12} = -\vec{F}_{21}$.

Rewriting this expression as $\vec{F}_{12} + \vec{F}_{21} = 0$ leads to an interesting implication. Since each force is a measure of the rate of change of momentum for that object, we can write

$$\vec{F}_{12} = \frac{\Delta \vec{p}_1}{\Delta t} \quad \text{and} \quad \vec{F}_{21} = \frac{\Delta \vec{p}_2}{\Delta t}$$

Therefore:

$$\frac{\Delta \vec{p}_1}{\Delta t} + \frac{\Delta \vec{p}_2}{\Delta t} = 0 \quad \text{and} \quad \frac{\Delta(\vec{p}_1 + \vec{p}_2)}{\Delta t} = 0$$

The change in the sum of the momenta is therefore zero, implying that the sum of the total momentum for the system $(\vec{p}_1 + \vec{p}_2)$ is a constant all the time. This conclusion is called the **law of conservation of linear momentum**, and we say simply that the momentum is conserved.

Here is another way of writing this conservation statement in a general form for any two masses (after separating all initial and final terms):

$$m_1 \vec{v}_{1i} + m_2 \vec{v}_{2i} = m_1 \vec{v}_{1f} + m_2 \vec{v}_{2f}$$

Extension of the law of conservation of momentum to two or three dimensions involves the recognition that momentum is a vector quantity. Given two masses moving in a plane relative to a coordinate system, conservation of momentum must hold simultaneously in both the horizontal and vertical directions. These vector components of momentum can be calculated using the standard techniques of vector analysis used to resolve forces in components.

> **REMEMBER**
>
> In an isolated system, the total momentum before an interaction is equal to the total momentum after the interaction.

Elastic Collisions

During any collision between two pieces of matter, momentum is always conserved. This statement is not, however, necessarily true about kinetic energy. If two masses stick together after a collision, it is observed that the kinetic energy before the collision is not equal to the kinetic energy after it. If the kinetic energy is conserved as well as the momentum, the collision is described as **elastic**.

If we have an elastic collision, we write both conservation laws to get two equations involving the velocities of the masses before and after the collision:

$$m_1 \vec{v}_{1i} + m_2 \vec{v}_{2i} = m_1 \vec{v}_{1f} + m_2 \vec{v}_{2f}$$
$$\frac{1}{2} m_1 v_{1i}^2 + \frac{1}{2} m_2 v_{2i}^2 = \frac{1}{2} m_1 v_{1f}^2 + \frac{1}{2} m_2 v_{2f}^2$$

> **TIP**
> On the AP Physics 1 exam, you may be asked to set up, but not solve, a system of equations. Your understanding of physics is thus demonstrated without having to spend the time doing the algebra!

If we cancel out the factor (1/2) in the second equation, and collect the expressions for each mass on each side, we can rewrite the two equations as

$$m_1 (\vec{v}_{1i} - \vec{v}_{1f}) = m_2 (\vec{v}_{2f} - \vec{v}_{2i})$$
$$m_1 (\vec{v}_{1i}^2 - \vec{v}_{1f}^2) = m_2 (\vec{v}_{2f}^2 - \vec{v}_{2i}^2)$$

The second equation is factorable and so can be simplified. Again rewriting the expressions, we get

$$m_1 (\vec{v}_{1i} - \vec{v}_{1f}) = m_2 (\vec{v}_{2f} - \vec{v}_{2i})$$
$$m_1 (\vec{v}_{1i} + \vec{v}_{1f})(\vec{v}_{1i} - \vec{v}_{1f}) = m_2 (\vec{v}_{2f} + \vec{v}_{2i})(\vec{v}_{2f} - \vec{v}_{2i})$$

If we take the ratio of the two expressions, we arrive at an interesting result after collecting terms:

$$\vec{v}_{1i} - \vec{v}_{2i} = -(\vec{v}_{1f} - \vec{v}_{2f})$$

This expression states that the relative velocity between the masses before the elastic collision is equal and opposite to the relative velocity between the masses after the elastic collision! Thus, the characteristic "rebounding" can be observed during elastic collisions.

Sample Problem

A mass of 2 kg is moving at a speed of 10 m/s along a horizontal, frictionless surface. It collides, and sticks, with a 3-kg mass moving in the same direction at 5 m/s.

(a) What is the final velocity of the system after the collision?
(b) What percentage of kinetic energy was lost in the collision?

Solution

(a) Using conservation of momentum, we see that

$$m_1 \vec{v}_1 + m_2 \vec{v}_2 = (m_1 + m_2) \vec{v}_f$$
$$(2 \text{ kg})(10 \text{ m/s}) + (3 \text{ kg})(5 \text{ m/s}) = (5 \text{ kg}) \vec{v}_f$$
$$(35 \text{ kg} \cdot \text{m/s}) = (5 \text{ kg}) \vec{v}_f$$
$$\vec{v}_f = 7 \text{ m/s}$$

> **TIP**
> For elastic collisions, the kinetic energy of the system is conserved.

(b) The initial kinetic energy is given by

$$\frac{1}{2}m_1v_1^2 + \frac{1}{2}m_2v_2^2 = KE_i$$

$$\frac{1}{2}(2\text{ kg})(10\text{ m/s})^2 + \frac{1}{2}(3\text{ kg})(5\text{ m/s})^2 = 137.5\text{ J}$$

The final kinetic energy is given by

$$\frac{1}{2}(m_1+m_2)v_f^2 = KE_f$$

$$\frac{1}{2}(5\text{ kg})(7\text{ m/s})^2 = 122.5\text{ J}$$

We now take the ratio of the final kinetic energy and the initial kinetic energy:

$$(122.5\text{ J})/(137.5\text{ J}) = 0.89 \rightarrow 89\%\text{ left over } KE$$

Therefore, 11% of the initial kinetic energy was lost in the collision.

> **Sample Problem**

A car traveling at 15 m/s rear ends a car in front of it traveling at only 8 m/s. If this collision is approximated as elastic and the car that was hit from the rear is now traveling at 11 m/s, what is final velocity of the other car?

> **Solution**

Since the collision is elastic, the masses of the cars are not needed. The relative speeds of the two cars must be maintained:

$$\text{Initially, relative speed} = 15 - 8 = 7$$

They must be the same as negative relative velocities after the collision:

$$-(x - 11) = 7$$
$$x = 4$$

The car that was traveling 15 m/s is now only traveling 4 m/s.

Inelastic Collisions

If the kinetic energy is not conserved after the collision (e.g., energy being lost to heat or friction), the collision is described as **inelastic**. Generally speaking, most everyday collisions will involve some loss of kinetic energy during the collision and thus are inelastic in nature. A **perfectly inelastic** collision is one in which the two objects are stuck together as one combined mass after the collision.

As an example, suppose that a mass m has a velocity $\vec{v}$, while mass M is at rest along a horizontal, frictionless surface. The two masses collide and stick together. What is the final velocity, $\vec{u}$, of the system? According to the law of conservation of momentum, the total momentum before the collision $(m\vec{v})$ must be equal to the total momentum after. Since the two masses are combining, the new mass of the system is $M + m$, and the new momentum is given by $(m + M)\vec{u}$. Thus, we find that:

$$\vec{u} = \frac{m\vec{v}}{m+M}$$

The initial kinetic energy is $(1/2)mv^2$. The final kinetic energy is given by

$$KE_f = \frac{1}{2}(m+M)\vec{u}^2 = \frac{1}{2}\left(\frac{m^2\vec{v}^2}{m+M}\right)$$

which is, of course, not equal to the initial kinetic energy.

> **Sample Problem**

Assuming the two objects collide completely inelastically (they stick together), rank the magnitude of the momenta of the product of these collisions.

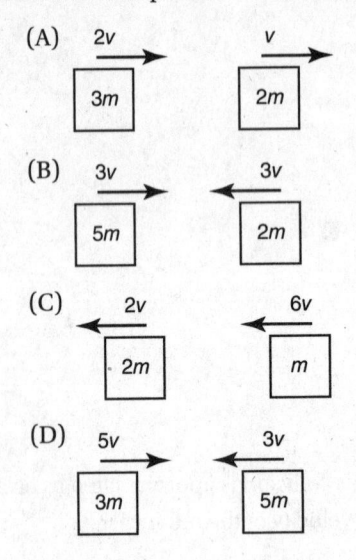

(A) →2v 3m   →v 2m

(B) →3v 5m   ←3v 2m

(C) ←2v 2m   ←6v m

(D) →5v 3m   ←3v 5m

Follow-up: Which collision will release the most energy?

✓ **Solution**

Simply sum the momenta of object A and object B before the collision. (Keep track of negative signs!) Since momentum is conserved, the highest magnitude (absolute value) total will be the highest after the collision as well.

(A) $6mv + 2mv = 8mv$
(B) $15mv - 6mv = 9mv$
(C) $-4mv - 6mv = -10mv$
(D) $15mv - 15mv = 0$

So, from highest magnitude of momentum to lowest:

$$C \to B \to A \to D$$

Follow-up answer: Crash D has the greatest loss of kinetic energy since all of the kinetic energy is lost during its collision. So crash D will release the greatest amount of energy.

Center of Mass

The center of mass of a system is a very useful concept. Conceptually, the center of mass is the point in a system about which all the mass is balanced. It is found by taking the weighted average of all points of mass along each axis. For example, along the x-axis:

$$x_{cm} = \frac{\sum m_i x_i}{\sum m_i}$$

It is the center of mass of an object to which we apply our forces in our free-body diagrams. The center of mass of an object is actually located at the positions specified in all our physics problems. For example, a real-world, unsymmetrical object may not appear to travel in a parabola when undergoing projectile motion as it twists and turns during its flight. However, its center of mass is following the parabolic path.

TIP

Note that the AP Physics 1 exam will not ask you to calculate the center of mass for more than 5 particles in two dimensions.

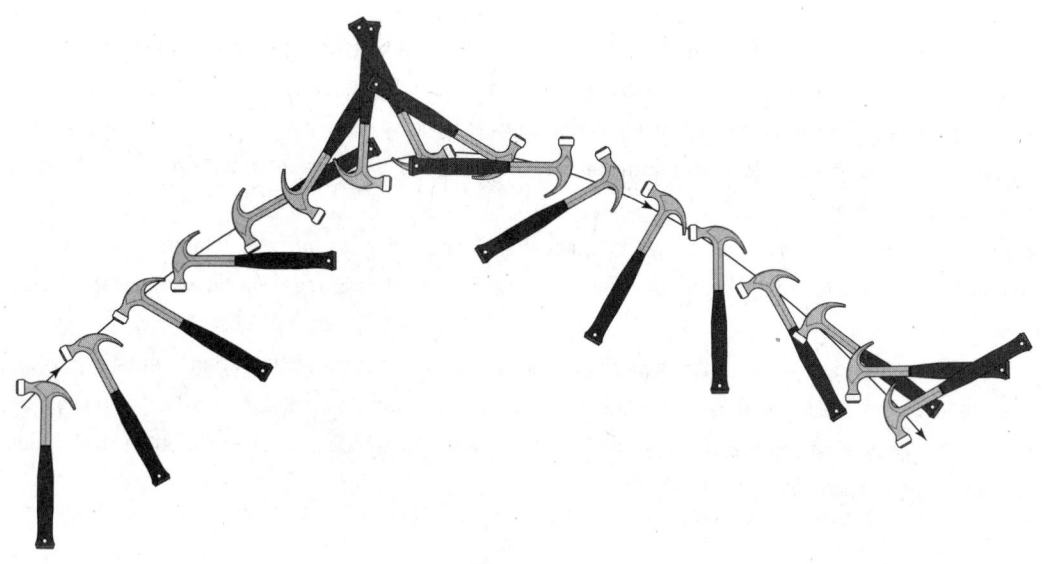

Figure 6.3

When thinking of the momentum of a system of interacting particles, the overall momentum of the system obeys all of Newton's laws relative to external forces. For example, a mass of gas consisting of many moving and interacting molecules can be modeled as being attracted gravitationally to Earth by placing the total mass of the gas cloud at its center of mass. Specifically, a system of interacting particles can be modeled as a single object located at its center of mass and having a velocity defined by

$$\vec{v}_{cm} = \frac{\sum_i \vec{p}_i}{\sum m_i} = \frac{\sum_i (m_i \vec{v}_i)}{\sum m_i}$$

which, in turn, obeys all of Newton's laws.

SUMMARY

- Momentum is a vector quantity equal to $m\vec{v}$.
- The impulse ($\vec{F}\Delta t$) is found by taking the area under a graph of force versus time.
- Impulse is equal to the change in momentum ($\vec{F}\Delta t = m\Delta\vec{v}$).
- Force is equal to the rate of change of momentum.
- In an isolated system, the total momentum is conserved.
- In an elastic collision, the kinetic energy of the system is conserved.
- In an inelastic collision (typically, but not necessarily) where masses stick together, kinetic energy is lost (this energy loss typically transforms into heat).

Problem-Solving Strategies for Impacts and Linear Momentum

In any isolated system, momentum is always conserved. Therefore, when you read a problem that does not state explicitly that momentum is involved, you can safely assume, in the case of collision or an impact, that total momentum is conserved, and you should write the equations for the conservation of momentum. The kinetic energy, however, is not necessarily conserved unless the collision is elastic. In summary, you should:

1. Decide whether an impact or a collision is involved. If there is a collision, observe whether it is elastic or inelastic.
2. Remember that, if the collision is inelastic, masses will usually stick together, so be sure to determine the new combined mass.

3. If the collision is elastic, write the equations for the conservation of both momentum and kinetic energy.
4. Remember that the impulse given to a mass is equal to its change in momentum. The change in momentum is a vector quantity in the same direction as the impulse or net force.
5. Be sure to take into account algebraically any reversal of directions, and remember the sign conventions for left, right, up, and down motions.
6. Keep in mind that motions may be simpler if studied in the center-of-mass frame of reference. In such a frame, one considers the motion of the center of mass, as well as the motions of mass particles relative to the center of mass.
7. Remember that in two dimensions the center of mass will follow a smooth path after an internal explosion since the forces involved were internal and the initial momentum in that frame was zero. For example, if a projectile launched at an angle experiences an explosion mid-flight, the center of mass of all the debris still follows the original parabolic trajectory.

Practice Exercises

Multiple-Choice

1. Which of the following expressions, where p represents the linear momentum of the particle, is equivalent to the kinetic energy of a moving particle?

    (A) mp^2
    (B) $m^2/2p$
    (C) $2p/m$
    (D) $p^2/2m$

2. Two carts having masses 1.5 kg and 0.7 kg, respectively, are initially at rest and are held together by a compressed massless spring. When released, the 1.5-kg cart moves to the left with a velocity of 7 m/s. What is the velocity and direction of the 0.7-kg cart?

    (A) 15 m/s right
    (B) 15 m/s left
    (C) 7 m/s left
    (D) 7 m/s right

3. The product of an object's instantaneous momentum and its acceleration is equal to its

    (A) applied force
    (B) kinetic energy
    (C) power output
    (D) net force

4. A ball with a mass of 0.15 kg has a velocity of 5 m/s. It strikes a wall perpendicularly and bounces off straight back with a velocity of 3 m/s. The ball underwent a change in momentum equal to

    (A) 0.30 kg · m/s
    (B) 1.20 kg · m/s
    (C) 0.15 kg · m/s
    (D) 5 kg · m/s

5. What braking force is supplied to a 3,000-kg car traveling with a velocity of 35 m/s that is stopped in 12 s?

    (A) 29,400 N
    (B) 3,000 N
    (C) 8,750 N
    (D) 105,000 N

6. A 0.1-kg baseball is thrown with a velocity of 35 m/s. The batter hits it straight back with a velocity of 60 m/s. What is the magnitude of the average impulse exerted on the ball by the bat?

    (A) 3.5 N · s
    (B) 2.5 N · s
    (C) 7.5 N · s
    (D) 9.5 N · s

7. A 1-kg object is moving with a velocity of 6 m/s to the right. It collides and sticks to a 2-kg object moving with a velocity of 3 m/s in the same direction. How much kinetic energy was lost in the collision?

    (A) 1.5 J
    (B) 2 J
    (C) 2.5 J
    (D) 3 J

8. A 2-kg mass moving with a velocity of 7 m/s collides elastically with a 4-kg mass moving in the opposite direction at 4 m/s. The 2-kg mass reverses direction after the collision and has a new velocity of 3 m/s. What is the new velocity of the 4-kg mass?

    (A) −1 m/s
    (B) 1 m/s
    (C) 6 m/s
    (D) 4 m/s

9. A mass m is attached to a massless spring with a force constant k. The mass rests on a horizontal, frictionless surface. The system is compressed a distance x from the spring's initial position and then released. The momentum of the mass when the spring passes its equilibrium position is given by

    (A) $x\sqrt{mk}$
    (B) $x\sqrt{k/m}$
    (C) $x\sqrt{m/k}$
    (D) $x\sqrt{k^2 m}$

10. During an inelastic collision between two balls, which of the following statements is correct?

    (A) Both momentum and kinetic energy are conserved.
    (B) Momentum is conserved, but kinetic energy is not conserved.
    (C) Momentum is not conserved, but kinetic energy is conserved.
    (D) Neither momentum nor kinetic energy is conserved.

Additional Practice

1. Two blocks with masses 1 kg and 4 kg, respectively, are moving on a horizontal, frictionless surface. The 1-kg block has a velocity of 12 m/s, and the 4-kg block is ahead of it, moving at 4 m/s, as shown in the diagram below. The 4-kg block has a massless spring attached to the end facing the 1-kg block. The spring has a force constant k equal to 1,000 N/m.

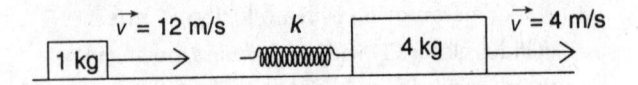

    (a) What is the maximum compression of the spring after the collision?
    (b) What are the final velocities of the blocks after the collision has taken place?

2. A 0.4-kg disk is initially at rest on a frictionless, horizontal surface. It is hit by a 0.1-kg disk moving horizontally with a velocity of 4 m/s. After the collision, the 0.1-kg disk has a velocity of 2 m/s at an angle of 43° to the positive x-axis.

    (a) Determine the velocity and direction of the 0.4-kg disk after the collision.
    (b) Determine the amount of kinetic energy lost in the collision.

3. A 50-kg girl stands on a platform with wheels on a frictionless, horizontal surface as shown below. The platform has a total mass of 1,000 kg and is attached to a massless spring with a force constant k equal to 1,000 N/m. The girl throws a 1-kg ball with an initial velocity of 35 m/s at an angle of 30° to the horizontal.

    (a) What is the recoil velocity of the platform-girl system?
    (b) What is the elongation of the spring?

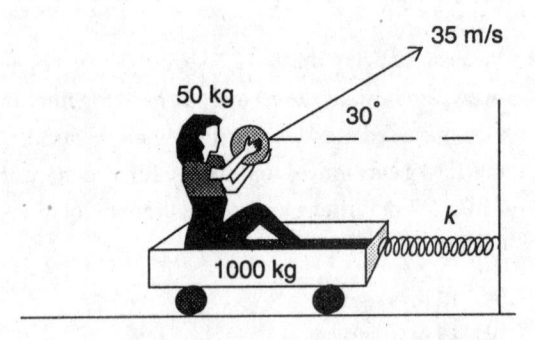

4. (a) Can an object have energy without having momentum? Explain.
   (b) Can an object have momentum without having energy? Explain.

5. Explain why there is more danger when you fall and bounce as opposed to falling without bouncing.

6. A cart of mass M is moving with a constant velocity v to the right. A mass m is dropped vertically onto it, and it is observed that the new velocity is less than the original velocity. Explain what has happened in terms of energy, forces, and conservation of momentum (as viewed from different frames of reference).

Answers Explained

Multiple-Choice Problems

1. **(D)** If we multiply the formula for kinetic energy by the ratio m/m, we see that the formula for kinetic energy becomes
$$KE = \left(\frac{1}{2m}\right)(m^2)(v^2) = \frac{p^2}{2m}$$

2. **(A)** Momentum is conserved, so $(1.5)(7) = 0.7v$. Thus, $v = 15$ m/s. The direction is to the right since in a recoil the masses go in opposite directions.

3. **(C)** An object's instantaneous momentum times its acceleration will equal its power output in units of joules per second or watts.

4. **(B)** The change in momentum is a vector quantity. The rebound velocity is in the opposite direction, so $\Delta v = 5 - (-3) = 8$ m/s. The change in momentum is
$$\Delta p = (0.15)(8) = 1.20 \text{ kg} \cdot \text{m/s}$$

5. **(C)** The formula is $F\Delta t = m\Delta v$. Solving for the force, we get
$$F = \frac{(3{,}000)(35)}{12} = 8{,}750 \text{ N}$$

6. **(D)** Impulse is equal to the change in momentum, which is
$$\Delta p = |0.1[-60 - 35]| = 9.5 \text{ N} \cdot \text{s}$$
because of the change in direction of the ball.

7. **(D)** First, we find the final velocity of this inelastic collision. Momentum is conserved, so we can write $(1)(6) + (2)(3) = (3)v'$ since both objects are moving in the same direction. Thus, $v' = 4$ m/s. The initial kinetic energy of the 1-kg object is 18 J, while the initial kinetic energy of the 2-kg mass is 9 J. Thus, the total initial kinetic energy is 27 J. After the collision, the combined 3-kg object has a velocity of 4 m/s and a final kinetic energy of 24 J. Thus, 3 J of kinetic energy has been lost.

8. **(B)** Momentum is conserved in this elastic collision but the directions are opposite, so we must be careful with negative signs. We therefore write $(2)(7) - (4)(4) = -(2)(3) + 4v'$ and get $v' = 1$ m/s.

9. **(A)** We set the two energy equations equal to solve for the velocity at the equilibrium position. Thus
$$\frac{1}{2}kx^2 = \frac{1}{2}mv^2$$
since no gravitational potential energy is involved, and we write that the velocity is $v = x\sqrt{k/m}$. Now momentum $p = mv$, so we multiply by m and factor the "mass" back under the radical sign, where it is squared so that we get $p = x\sqrt{mk}$.

10. **(B)** Momentum is always conserved in an inelastic collision, and kinetic energy is not conserved because the objects stick together.

Additional Practice

1. (a) Upon impact, the spring becomes compressed, but both blocks are still in motion. Therefore, for an instant, we have an inelastic collision, and momentum is, of course, conserved. Thus, we can write for the moment of impact
$$m_1\vec{v}_{1i} + m_2\vec{v}_{2i} = (m_1 + m_2)\vec{v}_f$$
Solving for the final velocity, we get $\vec{v}_f = 53.6$ m/s. Using the initial values for the velocities, we find that the initial kinetic energies are 72 J and 32 J for the 1-kg and 4-kg blocks, respectively. Thus, the total initial kinetic energy is 104 J. Using the final velocity of 5.6 m/s and the combined mass of 5 kg, we get a final kinetic energy of 78.4 J. The difference in kinetic energy of 25.6 J is used to compress the spring in this inelastic collision. Using the formula for the work done against a spring, $(1/2)kx^2$, we get $x = 0.05$ m for the maximum compression of the spring after the collision.

(b) After the collision has taken place, the two blocks again separate. If we treat the situation as elastic (since we assume that the work done to compress the spring will be used by the spring in rebounding), we can use the initial velocities as a "before" condition for momentum and kinetic energy. We then seek to solve for the two final velocities of the blocks after the rebound has taken place. To find both velocities, we need two equations, and we use the conservation of kinetic energy in this case to assist us. Thus, we write

$$m_1 \vec{v}_{1i} + m_2 \vec{v}_{2i} = m_1 \vec{v}_{1f} + m_2 \vec{v}_{2f}$$

and for the kinetic energy

$$\tfrac{1}{2} m_1 v_{1i}^2 + \tfrac{1}{2} m_2 v_{2i}^2 = \tfrac{1}{2} m_1 v_{1f}^2 + \tfrac{1}{2} m_2 v_{2f}^2$$

Substituting the known numbers, we obtain for the momentum $28 = v_{1f} + 4 v_{2f}$, and for the kinetic energy we get $208 = v_{1f}^2 + 4 v_{2f}^2$.

If we solve for v_{1f} from the momentum equation, square the result, and substitute into the kinetic energy equation, we obtain a factorable quadratic equation for v_{2f}:

$$v_{2f}^2 - 11.2 v_{2f} + 28.2 = (v_{2f} - 7.2)(v_{2f} - 4) = 0$$

Of the two choices for the second final velocity, only one provides a physically meaningful set of solutions. This is so because the first mass must rebound, and hence its final velocity must be negative. Our final answers are therefore $v_{2f} = 7.2$ m/s and $v_{1f} = -0.8$ m/s.

2. (a) This is a two-dimensional collision, and we consider the conservation of momentum in each direction, before and after. In the x-direction, we have only the initial momentum of the 0.1-kg disk with a velocity of 4 m/s. After the collision, the 0.1-kg disk has an x-component of momentum given by $(0.1)(2)\cos 43$.

The 0.4-kg disk had initially zero momentum and now has some unknown velocity at some unknown angle, θ, to the x-axis. Let us assume that the angle is below the x-axis, so that the x-component will be positive and the y-component negative. If our assumption is correct, we will get a positive answer for θ. A negative answer will let us know that the assumption is incorrect. The x-component of final momentum for the 0.4-kg disk is given by $(0.4)v_{2f}\cos\theta$, and the y-component by $-(0.4)v_{2f}\sin\theta$. We can therefore write for the x-direction:

$$(0.1)(4) + 0 = (0.1)(2)\cos 43 + (0.4)v_{2f}\cos\theta$$

and for the y-direction:

$$0 = (0.1)(2)\sin 43 - (0.4)v_{2f}\sin\theta$$

Solving for the angles and velocities in each, we get the following two equations:

$$v_{2f}\sin\theta = 0.34 \quad \text{and} \quad v_{2f}\cos\theta = 0.635$$

Taking the ratio gives $\tan\theta = 0.535$ and $\theta = 28°$. Substituting this angle gives us the final velocity for the 0.4-kg disk: 0.72 m/s.

(b) To find the loss of kinetic energy, we calculate the total initial and final kinetic energies. Using the given data, we find that the initial kinetic energy is 0.8 J. Using our final data, we get a total final kinetic energy of 0.3 J. Thus, 0.5 J of kinetic energy has been lost.

3. (a) In this problem, the x-component of the velocity provides the impulse to elongate the spring using recoil. Thus, we have that the velocity of the ball is $35\cos 30 = 30.31$ m/s. Using conservation of momentum, we state that $(1)(30.31) = (1,050)v'$. Thus, $v' = 0.0289$ m/s in the negative x-direction, so the recoil velocity of the platform-girl system can be expressed as $v' = -0.0289$ m/s.

$$\left(\tfrac{1}{2}\right)(1,050)(0.0289)^2 = \left(\tfrac{1}{2}\right)(1,000) x^2$$

(b) Solving for x gives $x = 0.0296$ m for the elongation of the spring.

4. (a) In a given frame of reference, an object can appear to be at rest and possess potential energy. In this frame, the object has zero velocity and hence its momentum is zero.

   (b) If an object has nonzero momentum, it must have nonzero velocity, which means it must have kinetic energy.

5. When an object bounces, an additional upward force (impulse) is given to it. This extra force, which provides for an additional change in momentum, can be dangerous if it is large enough.

6. First, if the mass is placed vertically onto the object, the increase in mass appears to lower the velocity since no force was acting in the direction of motion. We can also state that since the object did not have any horizontal motion, the friction between the new mass and the object must act to accelerate the mass m, and this energy comes from the moving object. From the mass M frame of reference, it appears as though the mass m is coming toward it, providing a backward force that will slow down the object.

7

Rotational Motion

Learning Objectives

In this chapter, you will learn about:
- → Parallel forces and moments
- → Torque
- → More static equilibrium problems using forces and torques
- → Rotational inertia
- → Angular kinematics
- → Energy considerations for rolling objects
- → Angular momentum

In Chapter 3, we discussed the fact that, if a single force is directed toward the center of mass of an extended rigid object, the result is a linear or translational acceleration in the same direction as the force. If, however, the force is not directed through the center of mass, then the result is a rotation about the center of mass. In addition, if the object is constrained by a fixed pivot point, the rotation will take place about that pivot (e.g., a hinge). These observations are summarized in Figure 7.1.

> **TIP**
> Torque produces rotation.

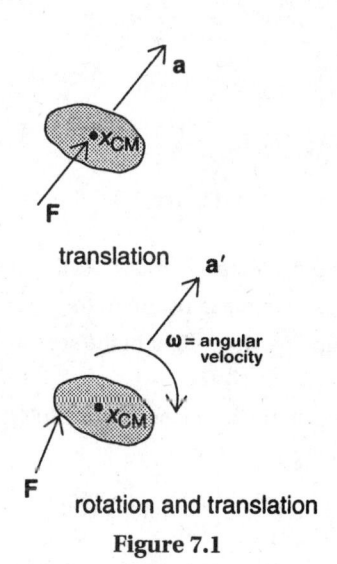

Figure 7.1

Parallel Forces and Moments

If two forces are used, it is possible to prevent the rotation if these forces are parallel and are of suitable magnitudes. If the object is free to move in space, translational motion may result (see Figure 7.2).

a
translation

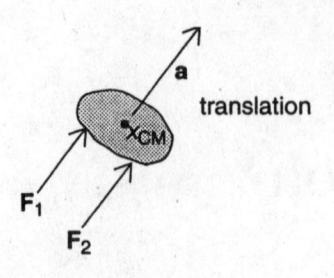

F₁
F₂

parallel forces
Figure 7.2

As an example of parallel forces, consider two people on a seesaw, as shown in Figure 7.3. The two people have weights F_{g1} and F_{g2}, respectively, and are sitting distances d_1 and d_2 from the fixed pivot point (called the **fulcrum**). From our discussion about forces, we see that the tendency of each force (provided by gravity) is to cause the seesaw to rotate about the fulcrum. Force F_{g1} will tend to cause a counterclockwise rotation; force F_{g2}, a clockwise rotation. Arbitrarily, we state that a clockwise rotation is taken as being a negative, while counterclockwise is taken as positive. What factors will influence the ability of the two people to remain in "balance"; that is, what conditions must be met so that the system remains in rotational equilibrium (it is already in a state of translational equilibrium and constrained to remain that way)?

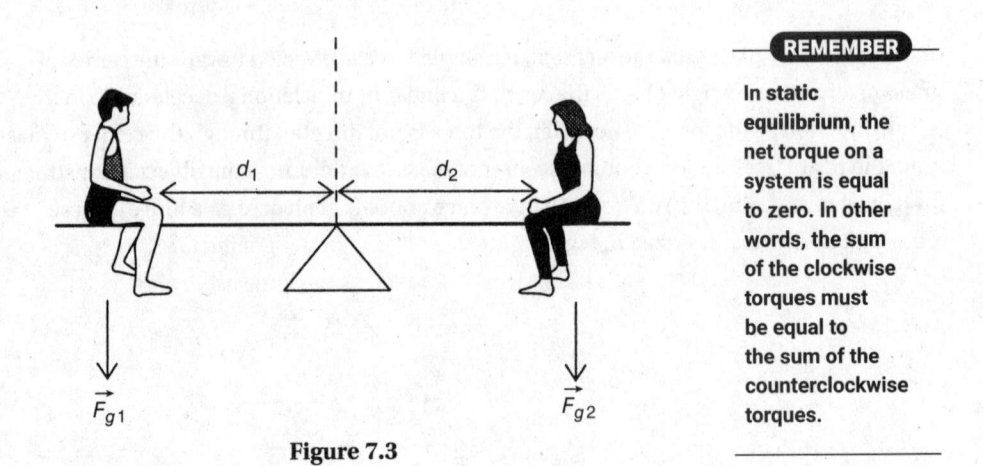

Figure 7.3

> **REMEMBER**
>
> In static equilibrium, the net torque on a system is equal to zero. In other words, the sum of the clockwise torques must be equal to the sum of the counterclockwise torques.

Through experiments, you can show that distances d_1 and d_2 (called **moment arm distances** or "lever arms") play a crucial role since we take the weights as being constant. If the two weights were equal, it should not be a surprise to learn that $d_1 = d_2$. In these examples, the weight of the seesaw is taken to be negligible (this is not a realistic scenario).

It turns out that, if F_{g2} is greater, that person must sit closer to the fulcrum. If equilibrium is to be maintained, the following condition must hold:

$$\left(F_{g_1}\right)\left(d_1\right) = \left(F_{g_2}\right)\left(d_2\right)$$

These are force and distance products, but they do not represent work in the translational sense because the force vectors are not causing a displacement. In rotational motion, the product of a force and a perpendicular moment arm distance (relative to a fulcrum) is called **torque**. Let us now consider some more examples of torques and equilibrium.

> **Sample Problem**

A 45-kg girl and a 65-kg boy are sitting on a seesaw in equilibrium. If the boy is sitting 0.7 m from the fulcrum, where is the girl sitting?

✓ **Solution**

In equilibrium,

$$F_{g_1} d_1 = F_{g_2} d_2$$
$$(65 \text{ kg})(9.8 \text{ m/s}^2)(0.7 \text{ m}) = (45 \text{ kg})(9.8 \text{ m/s}^2) d_2$$
$$d_2 = 1.01 \text{ m}$$

Torque

When you tighten a bolt with a wrench, you apply a force to create a rotation or twist. This twisting action, called torque in physics, is represented by the Greek letter τ. Even though its units are newton · meters (N · m), you must not confuse it with translational work. In static equilibrium, the two conditions met are that the vector sum of all forces acting on the object equals zero and that the vector sum of all torques equals zero:

$$\sum \vec{F} = 0 \quad \text{and} \quad \sum \vec{\tau} = 0$$

When we say that $\tau = Fd$, d is the moment arm distance from the center of mass, or the pivot point, and F is the force perpendicular to the vector displacement. If the force is applied at some angle θ to the object, as seen in Figure 7.4, the component of the force, perpendicular to the vector displacement out from the pivot, is taken as the force used. Algebraically this is stated as $F_\perp = F \sin \theta$.

REMEMBER

Torque is a vector quantity, but work is not! The direction of the torque is taken as either positive or negative, depending on whether the rotation is counterclockwise (+) or clockwise (−).

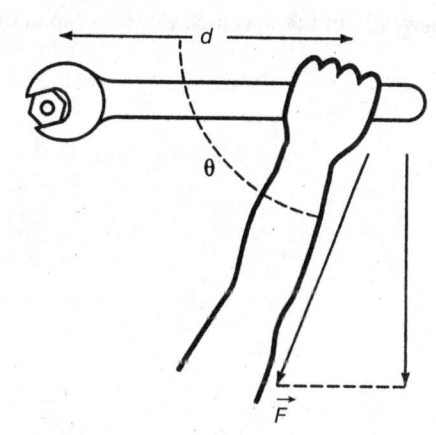

Figure 7.4

$$\tau = Fd \sin \theta$$

> **Sample Problem**

Consider a light string wound around a frictionless and massless wheel, as shown in Figure 7.5. The free end of the string is attached to a 1.2-kg mass that is allowed to fall freely. The wheel has a radius of 0.25 m. What torque is produced?

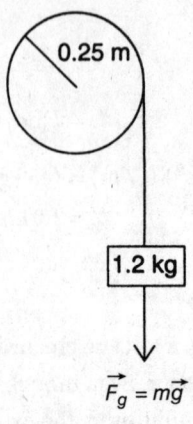

Figure 7.5

✓ **Solution**

The force acting at right angles to the center of the wheel is the weight of the mass, given by

$$F_g = mg = (1.2)(9.8) = 11.76 \text{ N}$$

The radius serves as the moment arm distance, $d = 0.25$ m, so we can write

$$\tau = -Fd = (11.76)(0.25) = -2.94 \text{ N} \cdot \text{m}$$

The torque is negative since the falling weight induces a clockwise rotation in this example.

> **Sample Problem**

Rank the magnitude of the net torque in each of the following situations. (All forces are equal in magnitude.) The axis of rotation is indicated with a dot.

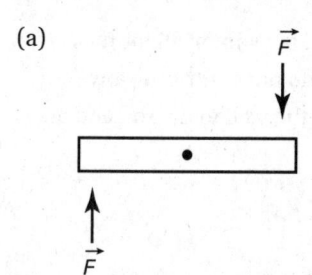

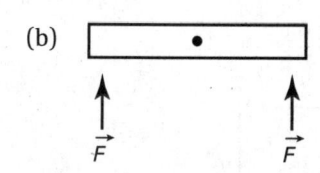

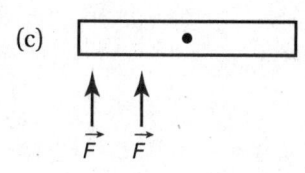

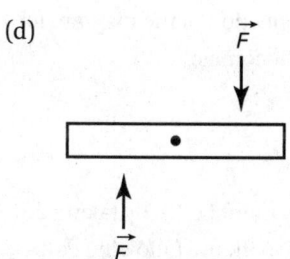

(a)

(b)

(c)

(d)

✓ **Solution**

Let r be the maximum lever arm (half the length of the rotating object). Using the convention of negative torque is clockwise torque, calculate the approximate net torque in each situation:

(a) $-rF - rF = -2rF$
(b) $+rF - rF = 0$
(c) $-rF - \left(\frac{r}{2}\right)F = -\left(\frac{3}{2}\right)rF$
(d) $-\left(\frac{r}{2}\right)F - \left(\frac{r}{2}\right)F = -rF$

Rank by magnitude (absolute value) from largest torque to smallest:
$$A \to C \to D \to B$$

More Static Equilibrium Problems Using Forces and Torques

In the following example (see Figure 7.6), a hinged rod of mass M is attached to a wall with a string (and, of course, by the hinge). The string is considered massless, and makes an angle θ with the horizontal. A mass m is attached to the end of the rod, whose length is ℓ. What is the tension T in the string?

Since the system is in static equilibrium, we must write that the sum of all forces and the sum of all torques equal zero. If we choose to focus on the pivot, then all forces acting through that point do not contribute any torques since the moment arm d would be zero. The reaction force $\vec{R}$ is the response of the wall to the rod and acts at some unknown angle ϕ. Thus, we state that

$$\sum \vec{\tau} = 0 \quad \text{and} \quad \sum \vec{F} = 0$$

Situation

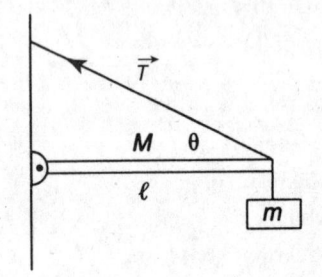

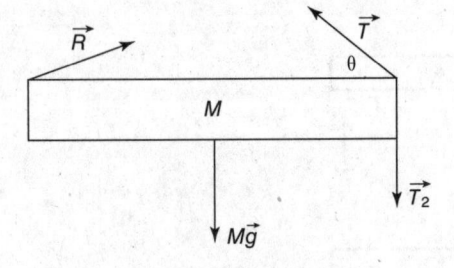

Free-body diagram for rod M

Figure 7.6

Since this is a static problem, both the net force and the net torques must be zero. By inspection of the diagram for hanging mass m, $T_2 = mg$. Thus, the equilibrium statement for vertical force on the rod becomes:

$$\sum F_x = R_x - T\cos\theta = 0$$
$$\sum F_y = R_y + T\sin\theta - Mg - mg = 0$$

The rod is assumed to be uniform. Therefore, its weight force is applied at the halfway point ($L/2$). By taking the left-hand side of the rod as our axis of rotation, the torque from force R is zero, leaving us with the following equilibrium statement for torques. The axis for a static problem can be taken anywhere you choose. By choosing the location where an unknown force is being applied, though, that unknown can be eliminated:

$$\sum \tau = (T\sin\theta)(L) - Mg(L/2) - mg(L) = 0$$

Note the clockwise torques are negative, while the counterclockwise torque is positive. Without additional information, the two force statements cannot be further simplified. However, the length of rod L can be eliminated from the torque equation. This allows us to solve for T in terms of the two masses:

$$T = ((M/2 + m)g)/\sin\theta$$

Usually, problems of this type will either give you the angle θ or let you determine the angle by the geometry of the situation. Once the tension $\vec{T}$ is known, one can determine the components of the unknown force $\vec{R}$ from the two force equations.

For example, if $M = 1$ kg, $m = 10$ kg, and $\theta = 30°$:

$$T = 205.8 \text{ N}$$
$$R_x = T\cos\theta = 178.2 \text{ N}$$
$$R_y = (M + m)_g - T\sin\theta = 4.9 \text{ N}$$

Note that the $\vec{R}$ force from the wall is almost entirely horizontal!

It should be noted that the rod is considered to have mass and that the torque produced by the rod is its weight taken from the center of mass ($\ell/2$ in this case) and is always clockwise (negative).

Rotational Inertia

Rotational inertia is slightly more complicated than linear inertia. However, like linear inertia, rotational inertia (or an object's moment of inertia) measures an object's tendency to resist changes to its motion, which is angular motion in this case. Linear inertia is simply the total mass of an object that we imagine as being located at the object's center of mass. Rotational inertia depends not only on the mass but also on how that mass is distributed. Specifically, for a distribution of point-like masses (or objects where the individual centers of mass are known), the rotational inertia (I) (i.e., the rotational inertia) of the entire object is given by:

$$I = \sum m_i r_i^2$$

where r is the distance from the rotational axis. The units are kg · m². The result depends not only on the shape and mass distribution of the object but also about which axis the object is being rotated. On the AP exam, if a solid object is being used, the general rotational inertia for that object will be given or easily obtained through another relationship (see below). For example, the rotational inertia of a solid rod of length L is different when rotating the object about one end of the rod as opposed to rotating it around its midpoint (see Figure 7.7).

> **TIP**
>
> The AP Physics 1 exam will not ask you to calculate moments of inertia for more than 5 objects in two dimensions. For any other shape, the momentum of inertia will be provided.

about midpoint: $mL^2/12$ about one end: $mL^2/3$

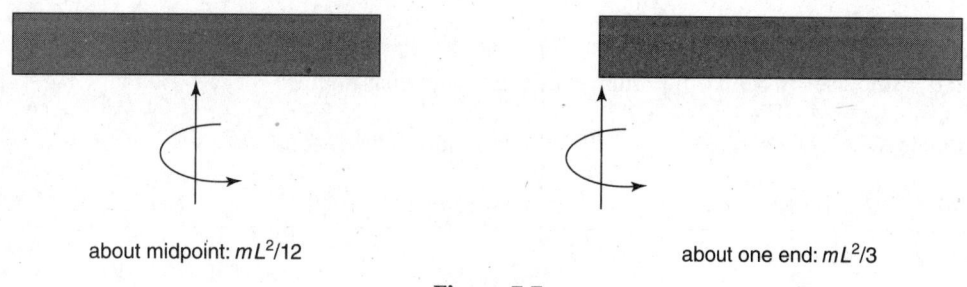

Figure 7.7

Try oscillating a ruler between your fingers while holding it at its midpoint. Then oscillate it while holding one end. You will quickly see that rotating the ruler about its midpoint is much easier and, hence, has a smaller rotational inertia. Generally speaking, one can adjust the rotational inertia of an object of mass m about its center of mass (I_{cm}) to be about a different but parallel axis via the **parallel axis theorem**:

$$I_{new} = I_{cm} + md^2$$

where d is the distance between the new axis of rotation and the center of mass.

Angular Kinematics

Rotational inertia takes the place of mass in all of the linear formulas in physics. An object's position in rotational motion is represented by its angular position (θ). Note that all angular values are measured in radians. (Recall the conversion 360 degrees = 2π radians = approximately 6.283 radians.) An object's angular velocity is defined analogously to linear velocity:

$$\omega = \frac{\Delta \theta}{\Delta t} \text{ (rad/sec)}$$

Likewise, angular acceleration (caused by torques) is defined kinematically by

$$\alpha = \frac{\Delta \omega}{\Delta t} \text{ (rad/s}^2)$$

Having defined the basic angular quantities of torque, rotational inertia, and the kinematic variables, we can complete the analogy between the physics of linear motion and that of rotational motion. Simply rewrite any of our linear physics relationships through substitution.

> **TIP**
> On the AP Physics 1 exam, all angular vector directions can be described simply as clockwise (CW) or counterclockwise (CCW).

Comparison of Linear and Angular/Rotational Motion

Linear	Angular/Rotational
x: position (m)	θ: angle (radians)
v: velocity (m/s)	ω: angular speed (rad/sec)
a: acceleration (m/s^2)	α: angular acceleration (rad/s^2)
$\Delta x = v_0 t + 1/2\, at^2$	$\Delta \theta = \omega_0 t + 1/2\, \alpha t^2$
F = force (push or pull) (newtons)	τ = torque = $|\vec{r}||\vec{F}|\sin\theta$
$\vec{F}_{net} = m\vec{a}$ (for constant m)	$\vec{\tau} = I\vec{\alpha}$ (for constant I)
$\vec{p} = m\vec{v}$ = momentum	$\vec{L} = I\vec{\omega}$ = angular momentum
$\Delta \vec{p}$ = impulse = $\vec{J} = \vec{F}\Delta t$	$\Delta \vec{L}$ = angular impulse = $\vec{\tau}\Delta t$
Linear $KE = 1/2\, mv^2$	Rotational $KE = 1/2\, I\omega^2$
Work = $|\vec{F}||\vec{D}|\cos\theta$	Work = $\tau \Delta \theta$

Sample Problem

If a net torque of 7.5 Nm is applied to an object initially at rest for 3.0 seconds, how fast will it be rotating? The object has a momentum of inertia of 15 kg m^2.

Solution

Determine the angular acceleration from the torque:

$$\vec{\tau} = I\vec{\alpha}$$
$$7.5 = 15\alpha$$
$$\alpha = 0.5 \text{ rad/s}^2$$

Use this angular acceleration in the equation of motion for angular speed:

$$\omega_f = \omega_i + \alpha_t$$
$$\omega_f = 0 + 0.5(3) = 1.5 \text{ rad/s}$$

Note that although radians (rad) are not technically a unit and are not formally required to be written down, it can be helpful to write them in to guide your thinking when solving the problem.

Energy Considerations for Rolling Objects

Generally speaking, objects that are rolling have both linear (sometimes called translational) kinetic energy $\left(\frac{1}{2}mv^2\right.$ using the velocity of the object's center of mass $\left.\right)$ and rotational kinetic energy $\left(\frac{1}{2}I\omega^2\right)$. If the cylindrical or spherical object is rolling without slipping, then the object's rotation speed (ω) and its center of mass speed (v) must be coupled by the radius of the object: $v = \omega r$.

Figure 7.8 shows various objects rolling without slipping down an incline.

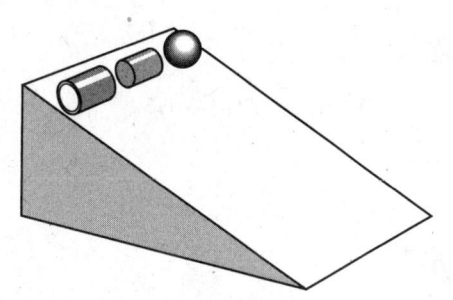

Figure 7.8

The mechanical energy of one of these objects is as follows: Mechanical energy = Gravitational PE + translational KE + rotational KE

$$mgh + \frac{1}{2}mv^2 + \frac{1}{2}I\omega^2$$

As usual with gravity problems, the mass cancels out since it is in every term. (Recall that rotational inertia, I, contains an m-term as well as an r^2-term.) Surprisingly, the rolling without slipping conditions cancels out the radius dependency in this expression as well (the r^2 in the momentum of inertia with the $\left(\frac{v}{r}\right)^2$ for ω). This means that the question of who will win the race comes down to the rotational inertia for each of the various objects. The object with the highest rotational inertia expression (not the actual value since the mass and radius cancel out) will require more of its mechanical energy for rotation. Thus, that object will have less of the total energy for the linear kinetic energy (representing the translational motion of the object). A solid sphere with a formula of $\frac{2}{5}mr^2$ is the winner in Figure 7.8! Note that rolling without slipping also insures that there is no mechanical energy lost via friction as the objects roll.

Angular Momentum

Just as forces cause changes in linear momentum over time, torques cause changes in angular momentum over time. In addition, an isolated system's angular momentum is conserved just as a linear momentum is. Whereas it is relatively rare for an isolated system's mass to change, changing the rotational inertia of an isolated object is not that difficult. If the momentum of inertia changes due to internal forces, the rotational velocity must also change in order to conserve angular momentum. Observe a spinning ice-skater drawing in her arms, thereby decreasing the r values in her rotational inertia calculation. You will see her rotational speed increase (see Figure 7.9).

For example, in the case of the solid sphere, the conservation of mechanical energy simplifies as follows:

$$E_{top} = E_{bottom}$$
$$mgh_i + 1/2\, mv_i^2 + 1/2\, I\omega_i^2 = mgh_f + 1/2\, mv_f^2 + 1/2\, I\omega_f^2$$

Rolling without slipping:

$$mgh_i + 1/2\, mv_i^2 + 1/2\, I(v_i/r)^2 = mgh_f + 1/2\, mv_f^2 + 1/2\, I(v_f/r)^2$$

Subbing in the expression for the rotational inertia:

$$mgh_i + 1/2\, mv_i^2 + 1/2\, (2/5mr^2)(v_i/r)^2 = mgh_f + 1/2\, mv_f^2 + 1/2\, (2/5mr^2)(v_f/r)^2$$

Canceling r's and m's:

$$gh_i + 1/2\, v_i^2 + 1/2\, (2/5)v_i^2 = gh_f + 1/2\, v_f^2 + 1/2\, (2/5)v_f^2$$

Starting from rest and setting final height to zero:

$$gh_i = 1/2\, v_f^2 + 1/5\, v_f^2$$

Figure 7.9

Arms and leg extend out — I_1, ω_1

Arms and leg withdraw inward — I_2, ω_2

$$I_1 \omega_1 = I_2 \omega_2$$

> ### Sample Problem

If the ice skater pictured in Figure 7.9 has decreased her original rotational inertia, I_1, by a factor of three, what happens to her original rotation rate of 2 turns per second?

✓ Solution

$$I_1 \omega_1 = I_2 \omega_2$$

since $I_2 = \frac{1}{3} I_1$.

Then, $\omega_2 = 3\omega_1$, to keep the equation balanced. Therefore, the final rotational rate will be 6 times per second.

> ### Sample Problem

Determine the angular momentum of Earth's orbital motion as it goes around the Sun.

Solution

Treating Earth as a point mass:

$$I = MR^2$$

where M is the mass of Earth and R is the orbital radius of the Earth about the Sun.

Next, determine the angular speed of Earth moving in a circle about the Sun:

$$\omega = v/R$$

where v is the orbital velocity of Earth.

Putting these expressions together into the angular momentum formula:

$$L = I\omega = MR^2(v/R)$$

Note that this is a general solution for the angular momentum of any point-like mass undergoing uniform circular motion!

Orbital speed of the Earth:

$$v = 2\pi R/(365 \text{ days}) \times (1 \text{ day})/(24 \text{ hours}) \times (1 \text{ hour}/3{,}600 \text{ s}) = 2.99 \times 10^4 \text{ m}$$
$$(R = 1.5 \times 10^{11} \text{ m})$$

(Note you can get the same result by using Newton's orbital speed equation and the mass of the Sun!)

$$L_{\text{Earth}} = MvR = (6.0 \times 10^{24})(2.99 \times 10^4)(1.5 \times 10^{11}) = 2.7 \times 10^{40} \text{ kgm}^2/\text{s}$$

SUMMARY

- Torque is a vector quantity equal to the product of (force × lever arm).
- The force and lever arm are perpendicular to each other.
- In equilibrium, the sum of the clockwise torques must be equal to the sum of the counterclockwise torques.
- Newton's laws apply in rotating systems just as they do in linear systems. Everything in the rotational world has an analogous equation in the linear.
- The units for torque are N · m, but they should *not* be confused with the scalar energy unit for joules, which can also be expressed as N · m.
- Angular momentum is conserved in isolated systems.

Problem-Solving Strategies for Torque

Solving torque problems is similar to solving static equilibrium problems with Newton's laws. In fact, Newton's laws provide the first condition for static equilibrium (the vector sum of all forces equals zero). This fact means that you should once again draw a free-body diagram of the situation.

In rotational static equilibrium, the vector sum of all torques must equal zero. Remember that torque is a vector even though its units are newton · meters. We take clockwise torques as negative and counterclockwise torques as positive. Any force going through the chosen pivot point (or fulcrum) does not contribute any torques. Therefore, it is wise to choose a point that eliminates the greatest number of forces. Also, remember that only the components of forces that are perpendicular to the direction of a radius displacement from the pivot (called the moment arm) are responsible, a situation that usually involves the sine of the force's orientation angle.

Problem-Solving Strategies for Rotational Motion

Translating a rotational problem first into the more comfortable language of linear physics can be helpful. The only major difference is that whereas changes of an object's mass in the middle of a problem are rare in linear problems, changes in rotational inertia are common in rotational problems.

Practice Exercises

Multiple-Choice

1. A 45-kg girl is sitting on a seesaw 0.6 m from the balance point, as shown below. How far, on the other side, should a 60-kg boy sit so that the seesaw will remain in balance?

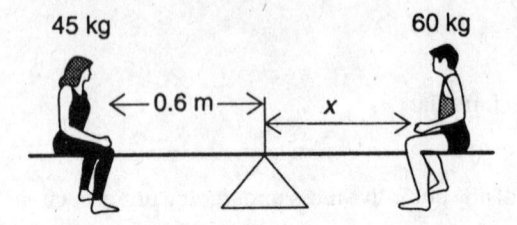

    (A) 0.30 m
    (B) 0.35 m
    (C) 0.40 m
    (D) 0.45 m

2. A balanced stick is shown below. The distance from the fulcrum is shown for each mass except the 10-g mass. What is the approximate position of the 10-g mass, based on the diagram?

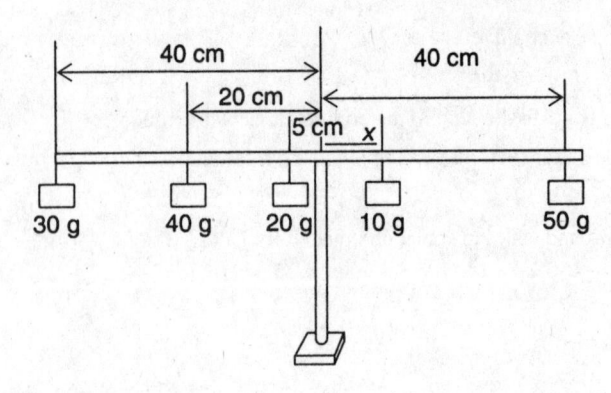

    (A) 7 cm
    (B) 9 cm
    (C) 10 cm
    (D) 15 cm

3. A solid cylinder consisting of an outer radius r_1 and an inner radius r_2 is pivoted on a frictionless axle as shown below. A string is wound around the outer radius and is pulled to the right with a force $F_1 = 3$ N. A second string is wound around the inner radius and is pulled down with a force $F_2 = 5$ N. If $r_1 = 0.75$ m and $r_2 = 0.35$ m, what is the net torque acting on the cylinder?

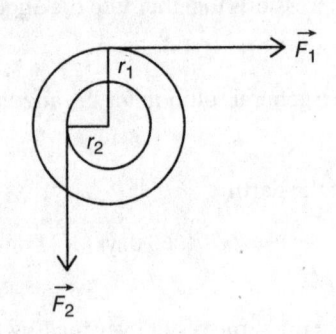

    (A) 2.25 N · m
    (B) −2.25 N · m
    (C) 0.5 N · m
    (D) −0.5 N · m

Answer questions 4 and 5 based on the following diagram. The rod is considered massless.

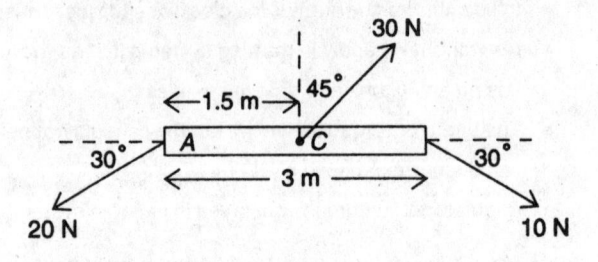

4. What is the net torque about an axis through point A?

    (A) 16.8 N · m
    (B) 15.2 N · m
    (C) −5.5 N · m
    (D) −7.8 N · m

5. What is the net torque about an axis through point C?

   (A) 3.5 N·m
   (B) 7.5 N·m
   (C) −15.2 N·m
   (D) 5.9 N·m

6. Compute the average angular acceleration and the angular displacement during the 2 seconds a rotating object speeds up from 0.5 rad/s to 0.7 rad/s.

   (A) $\alpha = 0.1$ rad/s² $\Delta\theta = 0.3$ radians
   (B) $\alpha = 0.2$ rad/s² $\Delta\theta = 0.5$ radians
   (C) $\alpha = 0.1$ rad/s² $\Delta\theta = 1.2$ radians
   (D) $\alpha = 0.2$ rad/s² $\Delta\theta = 1.2$ radians

7. An object with a rotational inertia I experiences a net torque of T. If the object is initially at rest, what is its angular speed after 3 seconds of this torque?

   (A) $3IT$
   (B) $9IT$
   (C) $3I/T$
   (D) $3T/I$

8. If a spinning ball of clay on top of a freely turning frictionless tabletop increases its rotational inertia by 50 percent by bulging outward, what will happen to the rotational speed of the clay and tabletop?

   (A) It will be half as fast.
   (B) It will be 1.5 times faster.
   (C) It will be 4 times slower.
   (D) It will be slower by a factor of 2/3.

Additional Practice

1. A 500-N person stands 2.5 m from a wall against which a horizontal beam is attached. The beam is 6 m long and weighs 200 N (see diagram below). A cable attached to the free end of the beam makes an angle of 45° to the horizontal and is attached to the wall.

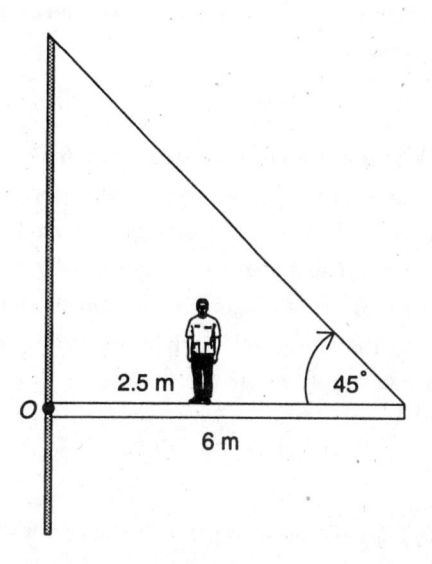

   (a) Draw a free-body diagram of the beam.
   (b) Determine the magnitude of the tension in the cable.
   (c) Determine the reaction force that the wall exerts on the beam.

2. A uniform ladder of length ℓ and weight 100 N rests against a smooth vertical wall. The coefficient of static friction between the bottom of the ladder and the floor is 0.5.

   (a) Draw a free-body diagram of this situation.
   (b) Find the minimum angle, θ, that the ladder can make with the floor so that the ladder will not slip.

3. A rotating toy top (rotational inertia = 0.012 kg m²) is rotating 8 times per second while encountering a friction torque of 0.2 Nm. For how much more time will it be rotating?

Answers Explained

Multiple-Choice Problems

1. **(D)** To remain in balance, the two torques must be equal. The force on each side is given by the weight, mg. The moment arm distances are 0.6 m and x. Since the factor g will appear on both sides of the torque balance equation, we can eliminate it and write

   $$(45)(0.6) = (60)x \quad \text{implies} \quad x = 0.45 \text{ m}$$

2. **(C)** In this problem, again, the sum of the torques on the left must equal the sum of the torques on the right. We could convert all masses to kilograms and all distances to meters, but in the balance equation the same factors appear on both sides. Therefore, for simplicity and time efficiency, we can simply write

   $$(30)(40) + (40)(20) + (20)(5) = 10x + (50)(40)$$
   $$x = 10 \text{ cm}$$

3. **(D)** The net torque is given by the vector sum of all torques. $\vec{F}_2$ provides a counterclockwise positive torque, while $\vec{F}_1$ provides a clockwise negative torque. Each radius is the necessary moment arm distance. Thus, we have

   $$\tau_{net} = (5)(0.35) - (3)(0.75) = -0.5 \text{ N} \cdot \text{m}$$

4. **(A)** In this problem, the net torque about point A implies that the force passing through point A does not contribute to the net torque. Also, we need the components of the remaining forces perpendicular to the beam. From the diagram, we see that the 30-N force acts counterclockwise (positive), while the 10-N force acts clockwise (negative). Thus:

   $$\tau_{net} = (30)(\cos 45)(1.5) - (10)(\sin 30)(3)$$
   $$= 16.8 \text{ N} \cdot \text{m}$$

5. **(B)** In this problem, since the pivot is now set at point C, we can eliminate the 30-N force passing through point C as a contributor to the torque. Again, we see that the 20-N force will act in a counterclockwise direction, while the 10-N force will act clockwise. Each force is 1.5 m from the pivot. We also need the component of each force perpendicular to the beam. Thus:

   $$\tau_{net} = (20)(\sin 30)(1.5) - (10)(\sin 30)(1.5)$$
   $$= 7.5 \text{ N} \cdot \text{m}$$

6. **(C)** $\alpha = \dfrac{\Delta \omega}{\Delta t} = \dfrac{0.7 - 0.5}{2} = 0.1 \text{ rad/s/s} = 0.1 \text{ rad/s}^2$

   $$\Delta \theta = \omega_0 t + \tfrac{1}{2}\alpha t^2 = 0.5(2) + \tfrac{1}{2}(0.1)(2)^2$$
   $$= 1 + 0.2$$
   $$= 1.2 \text{ radians}$$

7. **(D)** Since $\tau = I\alpha$ and $\alpha = \dfrac{\Delta \omega}{\Delta t}$, we can combine and solve for $\Delta \omega$:

   $$\Delta \omega = (T/I)\Delta t = 3T/I$$

8. **(D)** Angular momentum is conserved:

   $$I_i \omega_i = I_f \omega_f \quad I_f = (3/2) I_i$$
   $$I_i \omega_i = ((3/2) I_i) \omega_f$$
   $$\omega_f = 2/3 \, \omega_i$$

Additional Practice

1. (a) The free-body diagram for this situation appears below:

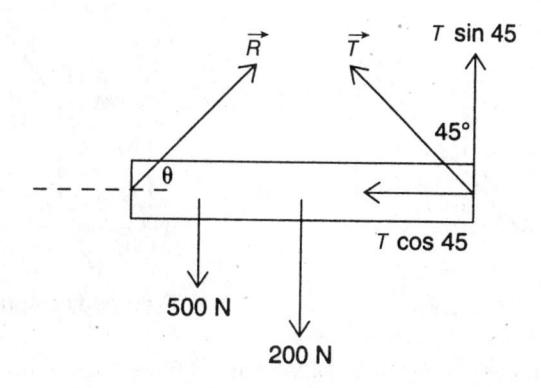

(b) The reaction force acts at some unknown angle, θ. We use our two conditions for equilibrium:

1. The sum of all forces in the x- and y-directions must equal zero:

$$0 = R\cos\theta - T\cos 45 \quad \text{and} \quad 0 = R\sin\theta + T\sin 45 - 500\text{ N} - 200\text{ N}$$

2. The sum of all torques through contact point O must be zero, thus eliminating the reaction force R:

$$0 = -(500)(2.5) - (200)(3) + T\sin 45(6)$$

Both weights produce clockwise torques! Solving for T in the last equation, we get $T = 436.05$ N.

(c) Using the result from part (b), we can rewrite the first two equations as:

$$R\cos\theta = (436.05)\cos 45 = 308.33\text{ N} \quad \text{and} \quad R\sin\theta = 391.67\text{ N}$$

Taking the ratio of these two equations gives us $\tan\theta = 1.27$; therefore, $\theta = 51.8°$. Since we know the angle at which the reaction force acts, we can simply write

$$R\cos 51.8 = 308.33 \quad \text{implies} \quad R = 498.6\text{ N}$$

2. (a) A sketch of the situation and its free-body diagram are given below:

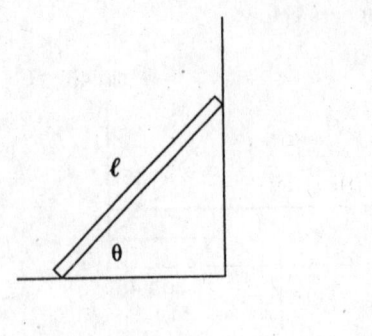

Situation 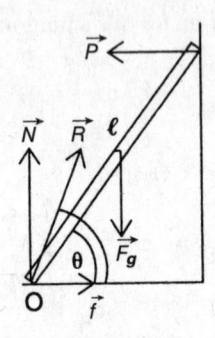   Free-body diagram

(b) The reaction force R is the vector resultant of the normal force N (from the floor) and the frictional force f that opposes slippage. The force P is the normal force of the vertical wall; there is no friction on the vertical wall. The weight F_g acts from the center of mass, $\ell/2$, and initiates a clockwise torque. Since we have equilibrium at this angle, we can write our equations for the conditions of static equilibrium.

   1. The sum of all x- and y-forces must be equal to zero. Horizontally, friction is an opposing force, as well as the reaction force, P: $0 = f - P$. Vertically, we can write $0 = N - F_g$. Since $F_g = 100$ N, then $N = 100$ N. Now, for no slippage, $f \leq \mu N$; thus, $f = (0.5)(100) = 50$ N $= P$.
   2. The sum of the torques must be zero. We need the component of P perpendicular to the ladder. From the geometry, we see that this is $P \sin \theta$. Also, the component of weight perpendicular to the ladder is $F_g \cos \theta$. Thus, for the sum of all torques equals zero, we write:

$$0 = P\ell \sin\theta - F_g\left(\frac{\ell}{2}\right)\cos\theta$$

The length of the ladder is therefore irrelevant and can be canceled out. Since we know that $P = 50$ N and $F_g = 100$ N, we find that $\tan\theta = F_g/2P = 1.0$. Therefore, $\theta = 45°$.

3. First use the given torque and momentum of inertia to determine the angular deceleration:

$$\vec{\tau} = I\vec{\alpha}$$
$$0.2 = (0.012)\alpha$$
$$\alpha = 16.7 \text{ rad/s}^2$$

An object has stopped rotating when its angular speed comes to zero ($\omega_f = 0$). Use the given rotation rate to determine the initial angular speed:

$$\omega_I = 8 \text{ rot/sec} \times (2\pi \text{ rad/rot}) = 16\pi \text{ rad/sec} = 50 \cdot \text{rad/s}$$

Use the angular deceleration from above in the equation of motion for angular speed:

$$\omega_f = \omega_i + \alpha t$$
$$0 = 50 - 16.7t$$
$$t = 3.0 \text{ s}$$

8
Oscillatory Motion

| **Learning Objectives**

In this chapter, you will learn about:
- → Simple harmonic motion: a mass on a spring
- → Simple harmonic motion: a simple pendulum
- → The dynamics of simple harmonic motion

Simple Harmonic Motion: A Mass on a Spring

From Hooke's law in Chapter 3, we know that a spring will become elongated by an amount directly proportional to the force applied. The force constant, k, relates to the specific amount of force (in newtons) needed to stretch or compress the spring by 1 meter:

$$\vec{F} = -k\vec{x}$$

The negative sign indicates that the force is restorative. One could easily use Hooke's law without the negative sign in the proper context.

Suppose we have a spring with a mass attached to it horizontally in such a way that the mass rests on a flat, frictionless surface as shown in Figure 8.1.

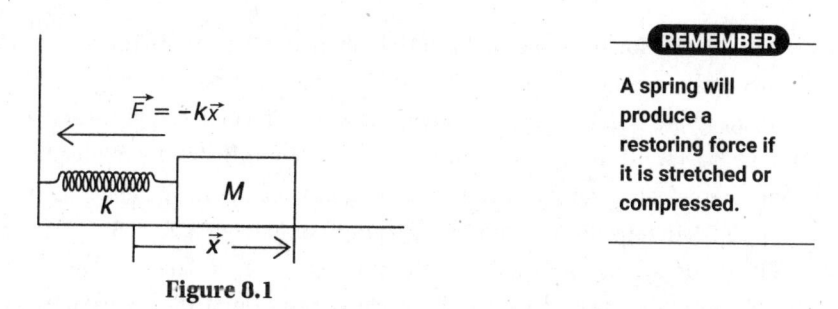

Figure 8.1

REMEMBER

A spring will produce a restoring force if it is stretched or compressed.

If the mass M is pulled a displacement $\vec{x}$, a restoring force of $\vec{F} = -k\vec{x}$ will act on it when released. However, as the mass accelerates past its equilibrium position, its momentum will cause it to keep going, thus compressing the spring. This action will slow the mass down until the same displacement, $\vec{x}$, is reached in compression. The same restoring force will then accelerate the mass back and forth, creating oscillatory motion with a certain period T (in seconds) and frequency f (in hertz or cycles per second). The SI unit for frequency is reciprocal seconds (s^{-1}).

Now, according to Newton's second law of motion, $\vec{F} = m\vec{a}$, so when the mass was originally extended, the restoring force $\vec{F} = -k\vec{x}$ would also produce an instantaneous acceleration, given by $\vec{F} = m\vec{a}$. In other words, $\vec{a} = -(k/m)\vec{x}$. The fact that the acceleration is directly proportional to the displacement (but in the opposite direction) is characteristic of a special kind of oscillatory motion called **simple harmonic motion**. From this expression, it can be shown that the acceleration is zero at the equilibrium point (where $x = 0$).

We can build up a qualitative picture of this type of motion by considering a displacement versus time graph for this mass-spring system. Suppose we have the system at rest so that the spring is unstretched. We now pull the mass to the right a distance A (called the **amplitude**) and release the mass. This action creates a restoring force that will pull the mass to the left toward the equilibrium point. The velocity will become greater and greater, reaching a maximum as the mass passes through the equilibrium point (at which $a = 0$). The mass will then move toward the left, slowing down as it compresses the spring (since the acceleration is in the opposite direction). The mass stops momentarily when $x = -A$ (because of conservation of energy) and then accelerates again to maintain simple harmonic motion (in the absence of friction).

A graph of displacement versus time (see Chapter 2) would look like a graph of the **cosine** function since the mass is beginning at some distance from the origin. We could consider the motion in progress from the point of view of the origin, in which case the graph would be of the **sine** function (some textbooks use this format). Figure 8.2 shows the characteristics of acceleration and deceleration as the mass oscillates in a period T with a frequency f (for arbitrary units of displacement and time).

From Chapter 4, we know that the maximum energy of a compressed (or stretched) spring is given by $E = (1/2)kA^2$, for $x = A$ as in our example. Thus, the constraining points $X = \pm A$ define the limits of oscillation for the mass.

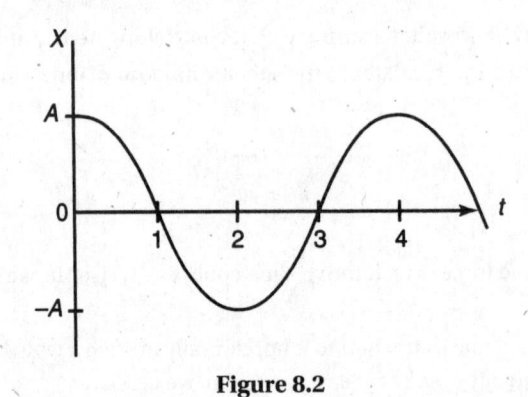

Figure 8.2

A graph of velocity versus time can be built up qualitatively in much the same way. Recall that the slope of the displacement versus time graph represents the instantaneous velocity. From Figure 8.2, you can see that, at $t = 0$, the graph is horizontal, indicating that $v = 0$. The increasingly negative slope shows that the mass is accelerating "backward" until, at $t = 1$, the line is momentarily straight, indicating maximum velocity when $x = 0$. The slope now gradually approaches zero at $t = 2$, indicating that the mass is slowing down as it approaches $x = -A$. The cycle then repeats itself, producing a graph similar to Figure 8.3.

Finally, analyzing the velocity graph with slopes, we produce an acceleration versus time graph. When $t = 0$, the line is momentarily straight with a negative slope, indicating a maximum negative acceleration. When the mass crosses the equilibrium point, velocity is maximum but acceleration is momentarily zero. The acceleration (proportional to displacement) reaches a maximum once again when $x = -A$. The cycle repeats itself, producing Figure 8.4. Notice that the acceleration graph is approximately shaped like the negative of the displacement graph, as expected!

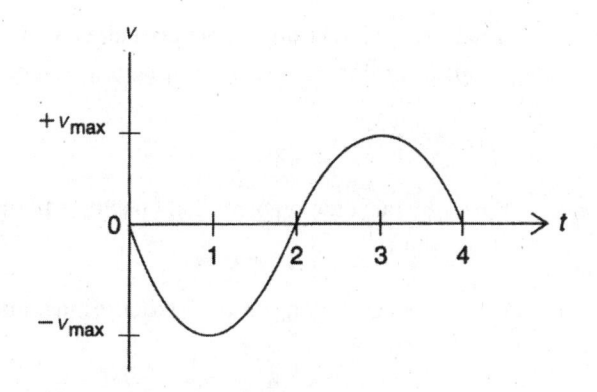

Figure 8.3

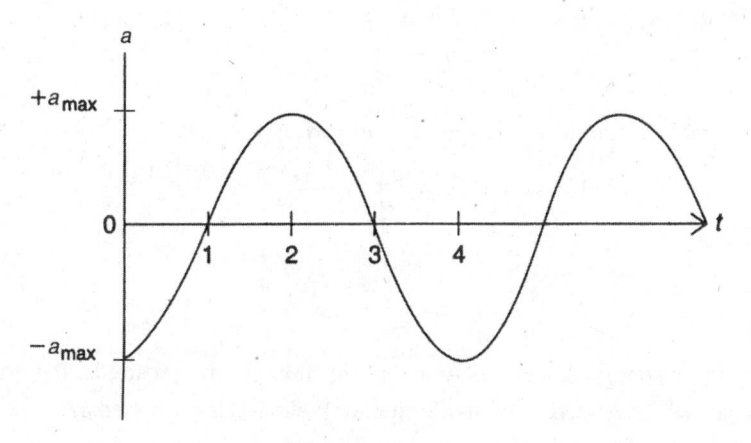

Figure 8.4

To derive the period of oscillation for the mass-spring system, we can consider another form of periodic motion already discussed: uniform circular motion. If a mass is attached to a rotating turntable and then turned onto its side, a projected shadow of the rotating mass simulates simple harmonic motion (Figure 8.5).

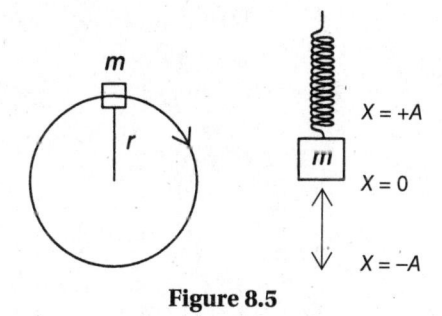

Figure 8.5

In this simulation, the radius r acts like the amplitude A, and the frequency and period of the rotation can be adjusted so that, when the shadow appears next to a real oscillating system, it is difficult to decide which one is actually rotating.

$$\omega = 2\pi f = 2\frac{\pi}{T}$$

Putting all this together, we can write a general equation of motion for simple harmonic motion:

$$x = A \cos 2\pi ft$$

From our understanding of uniform circular motion, we know that the magnitude of the centripetal acceleration is given by

$$a_c = \frac{4\pi^2 r}{T^2} = \frac{v^2}{r}$$

Since the projected sideways view of this motion appears to approximate simple harmonic motion, we can let the radius r be approximated by the linear displacement x and write

$$a_x = \frac{4\pi^2 x}{T^2}$$

Since $a = (k/m)x$ in simple harmonic motion, we can now write

$$\left(\frac{k}{m}\right)x = \frac{4\pi^2 x}{T^2}$$

$$\boxed{T = 2\pi \sqrt{\frac{m}{k}}}$$

> **TIP**
> The period of a mass on a spring is independent of the acceleration due to gravity.

This is the equation for the period of an oscillating mass-spring system in seconds. To find the frequency of oscillation, recall that $f = 1/T$. Also, the angular frequency (velocity) is expressed as $\omega = 2\pi/T$, which implies that $\omega = \sqrt{k/m}$. The units of angular frequency are radians per second. Note there is no dependence on amplitude.

> **Sample Problem**

A 0.5-kg mass is attached to a massless, elastic spring. The system is set into oscillation along a smooth, horizontal surface. If the observed period is 0.5 s, what is the value of the force constant k?

> **Solution**

The formula for the period is

$$T = 2\pi \sqrt{\frac{m}{k}}$$

Thus,

$$k = \frac{4\pi^2 m}{T^2}$$

$$k = \frac{(4\pi^2)(0.5 \text{ kg})}{(0.5 \text{ s})^2} = 79 \text{ N/m}$$

Simple Harmonic Motion: A Simple Pendulum

Imagine a pendulum consisting of a mass M (called a bob) and a string of length ℓ that is considered massless (Figure 8.6). The pendulum is displaced through an angle θ that is much less than 1 radian (about 57 degrees). Under these conditions, the pendulum approximates simple harmonic motion and the period of oscillation is independent of amplitude.

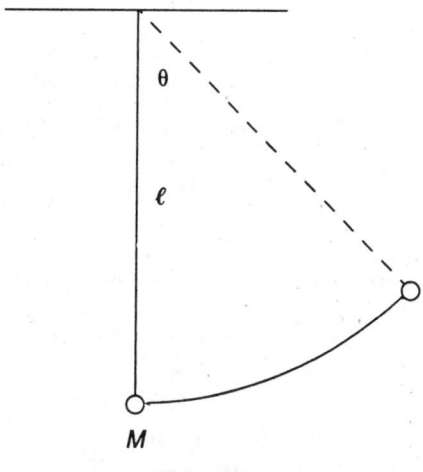

Figure 8.6

When the pendulum swings through an arc of length s, it appears to be following a straight path for a suitably chosen small period of time or small section of the arc. In this approximation, we can imagine the pendulum as being accelerated down an incline. The oscillations occur because gravity accelerates the pendulum back to its lowest position. Its momentum maintains the motion through that point, and then the constraining action of the string (providing a centripetal force) causes the pendulum to swing in an upward arc. Conservation of energy will bring the pendulum to the same vertical displacement (or cause it to swing through the same arc length) and then momentarily stop until gravity begins to pull it down again.

If we imagine that an incline of set angle θ is causing the acceleration, then, from our discussions of kinematics, we know that $a = -g \sin \theta$ (where the angle is measured in radians). Now, if θ is sufficiently less than 1 radian, we can write that $\sin \theta \approx \theta$ and therefore that

$$a = -g \sin \theta \approx -g\theta \approx \frac{gs}{\ell}$$

In the above equation we used the known relationship from trigonometry that, in radian measure, if s is the arc length and ℓ corresponds to the effective "radius" of swing, then $s = \ell\theta$.

Remember that these are only approximations, but for angles of less than 20 degrees, the approximations are fairly accurate. Since the pendulum now approximates simple harmonic motion, we know that we can find a suitable rotational motion that, when viewed in projection (as well as the pendulum swing being viewed in projection), simulates our simple harmonic motion. Therefore, in a similar fashion to that in the first section of this chapter, we can write

$$a = 4\frac{\pi^2 s}{T^2} = \frac{gs}{\ell}$$

where the arc length s is the amplitude of swing in this case.

If we solve for the period T, we finally get

$$T = 2\pi \sqrt{\frac{\ell}{g}}$$

TIP

The period of a simple pendulum is independent of the mass.

Notice that period is independent of mass but is very sensitive to the local acceleration of gravity. This becomes an excellent way to independently measure the value of g in various locations.

> **Sample Problem**

A simple pendulum consists of a string 0.4 m long and a 0.3-kg mass. On an unknown planet, the pendulum is set into oscillation and the observed period is 0.8 s. What is the value of g on this planet?

✓ **Solution**

The period of a simple pendulum is independent of the mass and is given by

$$T = 2\pi \sqrt{\frac{\ell}{g}}$$

Thus,

$$g = \frac{4\pi^2 \ell}{T^2} = \frac{(4\pi^2)(0.4 \text{ m})}{(0.8 \text{ s})^2} = 25 \text{ m/s}^2$$

The Dynamics of Simple Harmonic Motion

From Chapter 4 we know that the work done to compress a spring is also equal to the potential energy stored in the spring. This value is expressed as $(1/2)kx^2$. If the system is set into oscillation by stretching the spring by an amount $x = A$, the maximum energy possessed by the oscillating system is given by $U = (1/2)kA^2$. As the mass oscillates, it reaches maximum velocity when it passes through $x = 0$. At that point, the potential energy of the spring is zero, and since we are treating the cases without friction, the loss of potential energy is balanced by this gain in kinetic energy:

$$KE_{max} = \frac{1}{2}mv^2_{max} = \frac{1}{2}kA^2$$

If we want to treat cases in which the mass is between the two extremes of $x = 0$ and $x = \pm A$, we note that the spring will still possess some elastic potential energy (no gravitational potential energy is involved since the mass is oscillating horizontally). This implies that

$$\frac{1}{2}kA^2 = \frac{1}{2}mv^2 + \frac{1}{2}kx^2$$

If we solve for velocity, we get

$$v = \pm \sqrt{\frac{k}{m}(A^2 - x^2)}$$

REMEMBER

The period of a system oscillating with simple harmonic motion is independent of the amplitude. In *simple harmonic motion*, the acceleration is in the opposite direction and proportional to the displacement!

> **Sample Problem**

A mass oscillates about its equilibrium point from four different starting positions. Rank the speed of the oscillating mass from fastest speed to slowest speed. Justify your answer.

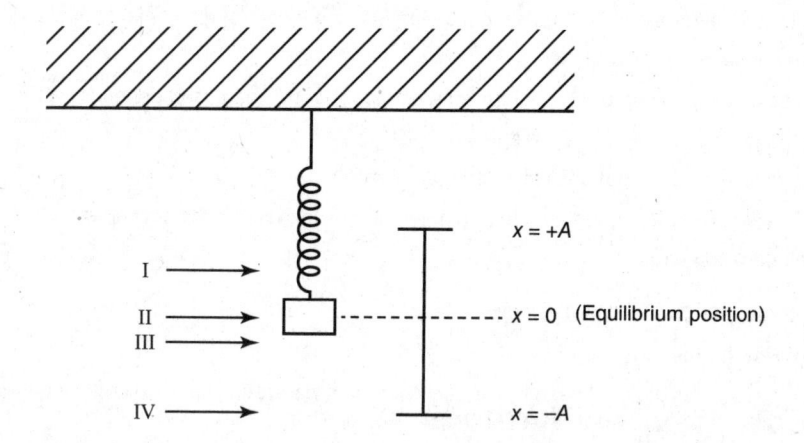

```
         ///////////////////
                 |
                 §
                 §         |─── x = +A
           I ──→ §
          II ──→ [ ]------- x = 0   (Equilibrium position)
         III ──→
                           |─── x = −A
          IV ──→
```

✓ **Solution**

From fastest to slowest speeds, the positions are ranked as follows:

$$II, III, I, IV$$

When the spring is fully compressed or fully extended (as in position IV), the velocity drops to zero as the mass is changing direction.

When the mass is traveling through the equilibrium point (as in position II), all of the energy is kinetic; therefore, the speed is the highest.

For positions I and III, the direction of motion does not matter since the question is asking about speed, not velocity. Since position III is closer to equilibrium, it has more kinetic energy than position I, making position III the slightly faster position.

SUMMARY

- Any periodic motion in which the acceleration is directly proportional to the negative of the displacement is called simple harmonic motion.
- The period of oscillation for a mass attached to a spring is independent of gravity and depends on the mass and force constant k.
- The period of oscillation for a simple pendulum is independent of the mass and depends on the length of the string and the value of the acceleration due to gravity.
- The period of simple harmonic motion is independent of amplitude.
- The velocity of motion is always greatest at the middle part of the motion. The acceleration is always greatest at the ends of the motion.

Problem-Solving Strategies for Oscillatory Motion

Keep in mind the following facts:

1. Simple harmonic motion is characterized by the fact that the acceleration varies directly with the displacement (but in the opposite direction).
2. A simple pendulum approximates simple harmonic motion only if its displacement angle is small ($\sim<15°$). The period will be independent of amplitude and mass under these small-angle circumstances.
3. It is sometimes easier to use energy considerations since the equations involve scalars and no free-body diagrams have to be drawn.
4. For pendulum or pendulum-like problems, if $\theta < 1$ radian, then $\sin\theta \approx \tan\theta \approx \theta$.

Practice Exercises

Multiple-Choice

1. What is the length of a pendulum whose period, at the equator, is 1 s?

   (A) 0.15 m
   (B) 0.25 m
   (C) 0.30 m
   (D) 0.45 m

2. On a planet, an astronaut determines the acceleration of gravity by means of a pendulum. She observes that the 1-m-long pendulum has a period of 1.5 s. The acceleration of gravity, in meters per second squared, on the planet is

   (A) 7.5
   (B) 15.2
   (C) 10.2
   (D) 17.5

3. When a 0.05-kg mass is attached to a vertical spring, it is observed that the spring stretches 0.03 m. The system is then placed horizontally on a frictionless surface and set into simple harmonic motion. What is the period of the oscillations?

   (A) 0.75 s
   (B) 0.12 s
   (C) 0.35 s
   (D) 1.3 s

4. A mass of 0.5 kg is connected to a massless spring with a force constant k of 50 N/m. The system is oscillating on a frictionless, horizontal surface. If the amplitude of the oscillations is 2 cm, the total energy of the system is

   (A) 0.01 J
   (B) 0.1 J
   (C) 0.5 J
   (D) 0.3 J

5. A mass of 0.3 kg is connected to a massless spring with a force constant k of 20 N/m. The system oscillates horizontally on a frictionless surface with an amplitude of 4 cm. What is the velocity of the mass when it is 2 cm from its equilibrium position?

   (A) 0.28 m/s
   (B) 0.08 m/s
   (C) 0.52 m/s
   (D) 0.15 m/s

6. If the length of a simple pendulum is doubled, its period will

   (A) decrease by 2
   (B) increase by 2
   (C) decrease by $\sqrt{2}$
   (D) increase by $\sqrt{2}$

7. A 2-kg mass is oscillating horizontally on a frictionless surface when attached to a spring. The total energy of the system is observed to be 10 J. If the mass is replaced by a 4-kg mass, but the amplitude of oscillations and the spring remain the same, the total energy of the system will be

   (A) 10 J
   (B) 5 J
   (C) 20 J
   (D) 3.3 J

Additional Practice

1. A mass M is attached to two springs with force constants k_1 and k_2, respectively. The mass can slide horizontally over a frictionless surface. Two arrangements, (a) and (b), for the mass and springs are shown below.

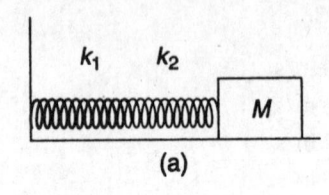

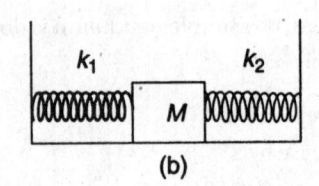

    (a) Show that the period of oscillation for situation (a) is given by
    $$T = 2\pi\sqrt{\frac{m(k_1 + k_2)}{k_1 k_2}}$$

    (b) Show that the period of oscillation for situation (b) is given by
    $$T = 2\pi\sqrt{\frac{m}{k_1 + k_2}}$$

    (c) Explain whether the effective spring constant for the system has increased or decreased.

2. A mass M is attached to two light elastic strings both having length ℓ and both made of the same material. The mass is displaced vertically upward by a small displacement Δy such that equal tensions T exist in the two strings, as shown below. The mass is released and begins to oscillate up and down. Assume that the displacement is small enough so that the tensions do not change appreciably. (Ignore gravitational effects.)

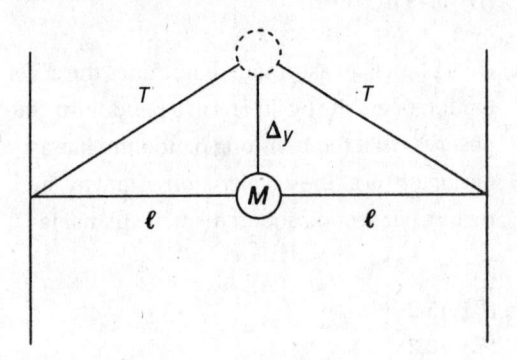

    (a) Show that the restoring force on the mass can be given by (for small angles)
    $$F = \frac{-2T\Delta y}{\ell}$$

    (b) Derive an expression for the frequency of oscillation.

3. Explain why a simple pendulum undergoes only approximately simple harmonic motion.

4. Explain how the mass of an object can be determined in a free-fall orbit using the concept of simple harmonic motion.

5. Both oscillating springs and oscillating pendula can be used as timekeeping devices. Compare the accuracy of both of these devices as timekeeping devices here on Earth versus on the Moon where gravity is 1/6 that of Earth's.

Answers Explained

Multiple-Choice Problems

1. **(B)** If we use the formula for the period of a pendulum, $T = 2\pi\sqrt{\ell/g}$, and square both sides, we can solve for the length:
$$\ell = T^2 \frac{g}{4\pi^2}$$
Using the values given in the question, we find that $\ell = 0.25$ m.

2. **(D)** Again using the formula for the period of a pendulum, this time we solve for the magnitude of the acceleration of gravity:
$$g = 4\frac{\pi\ell}{T^2}$$
Using the values given in the question, we find that $g = 17.5$ m/s².

3. **(C)** We first find the spring constant. From $F = kx$, we see that
$$F = mg = (0.05)(9.8) = 0.49 \text{ N}$$
Thus,
$$k = \frac{F}{x} = \frac{0.49}{0.03} = 16.3 \text{ N/m}$$
Now the period is given by the formula $T = 2\pi\sqrt{m/k}$. Using the values given in the question, we find that $T = 0.35$ s.

4. **(A)** The formula for total energy is $E = (1/2)kA^2$. Using the values given in the question, we get $E = 0.01$ J.

5. **(A)** The formula for any intermediary velocity is $v = \sqrt{(k/m)(A^2 - x^2)}$. Before substituting the known values, we must change centimeters to meters: 4 cm = 0.04 m and 2 cm = 0.02 m. Then, using these values for amplitude and position, we find that $v = 0.28$ m/s.

6. **(D)** Since the period of a pendulum varies directly as the square root of the length, if the length is doubled, the period increases by $\sqrt{2}$.

7. **(A)** The maximum energy in the mass-spring system can be expressed independently of the mass; it depends on the force constant and the square of the amplitude. Thus, if the spring and the amplitude of oscillations remain the same, so will the total energy of the system (10 J).

Additional Practice

1. (a) When mass M is stretched a distance x, spring k_1 is stretched distance x_1 and spring k_2 is stretched distance x_2. At the point of connection, Newton's third law states that the forces must be equal and opposite. Thus, we can write
$$k_1 x_1 = k_2 x_2$$
Since $x = x_1 + x_2$, we have $x_2 = x - x_1$, and so
$$k_1 x_1 = k_2(x - x_1)$$
$$= k_2 x - k_2 x_1$$
$$k_1 x_1 + k_2 x_1 = k_2 x$$
$$k_2 x = (k_1 + k_2) x_1$$
$$x_1 = \left(\frac{k_2}{k_1 + k_2}\right) x$$
is the expression for the displacement of k_1.

Now, for any spring, let's say spring 1:
$$F_1 = k_1 x_1 = \left(\frac{k_1 k_2}{k_1 + k_2}\right) x$$
Let's call $(k_1 k_2 / k_1 + k_2) = k'$, so
$$F_1 = k' x$$
This is in the form of the equation for simple harmonic motion and therefore is equal to ma by Newton's second law:
$$ma = k' x$$
Thus,
$$T = 2\pi\sqrt{\frac{m}{k'}} = 2\pi\sqrt{\frac{m(k_1 + k_2)}{k_1 k_2}}$$
This is the period of oscillation for arrangement (a).

(b) In this case, each spring is displaced the same distance as the mass, since the springs are on either side of the mass. Thus, we can write
$$F = -(k_1 + k_2) x_1$$
and set $k' = k_1 + k_2$. We therefore have the equation for the period of oscillation
$$T = 2\pi\sqrt{\frac{m}{k_1 + k_2}}$$
for arrangement (b).

(c) Situation (a) suggests that when the springs are connected in series, the effective spring constant decreases. Situation (b) suggests that when the springs are connected in parallel, the effective spring constant increases.

2. (a) In the diagram, we can see that, since Δy is small, $\theta = \Delta y/\ell$. Now, since θ (in radians) is small, $\theta \approx \sin\theta \approx \Delta y/\ell$. From the geometry, we see that the net restoring force is given by

$$\sum F = -2T\sin\theta = \frac{-2T\Delta y}{\ell}$$

(b) Since this situation is approximating simple harmonic motion, we can write

$$F = -ky$$
$$-2T\frac{\Delta y}{\ell} = -k\Delta y$$
$$k = 2\frac{T}{\ell}$$

so we must then have

$$f = \frac{1}{2\pi}\sqrt{\frac{2T}{\ell m}}$$

3. A pendulum represents only approximately simple harmonic motion since only for small angles is the acceleration proportional to the displacement (angular displacement in this case).

4. Since the period of a horizontally oscillating mass on a spring is independent of the acceleration due to gravity, the object's mass can be determined using springs and horizontal oscillations.

5. Examine our two formulas:

$$\text{Pendulum: } T = 2\pi\sqrt{\frac{\ell}{g}}$$
$$\text{Springs: } T = 2\pi\sqrt{\frac{m}{k}}$$

The pendulum will have a longer period (by a factor of $\sqrt{6}$) on the Moon since g will be six times smaller, whereas the spring's period will be unaffected. However, note that the equilibrium position for the hanging mass-on-a-spring system will be higher on the Moon as it will be stretched six times less by the weaker gravity there.

9

Fluids

Learning Objectives

In this chapter, you will learn about:
- → Static fluids
- → Pascal's principle
- → Static pressure and depth
- → Buoyancy and Archimedes' principle
- → Fluids in motion
- → Bernoulli's equation

Generally speaking, there are three everyday states of matter: solid, liquid, and gas. Solids are characterized by the fixed relative positions of the molecules and atoms within them and thus maintain their shape. Liquids are made of molecules or atoms with no fixed positions but still bonded together by intermolecular forces to have a definite volume but no fixed shape. Gases are made of relatively independent molecules and atoms and thus have a volume and shape defined by the container rather than by the gas itself.

Static Fluids

Fluids represent states of matter that take the shape of their containers. Liquids are referred to as *incompressible fluids*, while gases are referred to as *compressible fluids*. In a Newtonian sense, liquids do work by being displaced, while gases do work by compressing or expanding. As we shall see later, the compressibility of gases leads to other effects described by the subject of thermodynamics.

Fluids can exert pressure by virtue of their weight or force of motion. We have already defined the unit of pressure to be the **pascal**, which is equivalent to 1 N of force per square meter of surface area. An additional unit used in physics is the **bar**, where 1 bar = 100,000 Pa. Atmospheric pressure is sometimes measured in millibars.

$$\text{Pressure} = \frac{\text{Force}}{\text{Area}}$$

> **TIP**
> On the AP Physics 1 exam, ideal fluids are incompressible and have no viscosity.

Pascal's Principle

In a fluid, static pressure is exerted on the walls of the container. Within the fluid, these forces act perpendicular to the walls. If an external pressure is applied to the fluid, this pressure will be transmitted uniformly to all parts of the fluid. The last sentence is also known as **Pascal's principle** since it was developed by the French physicist Blaise Pascal.

Pascal's principle refers only to an external pressure. Within the fluid, the pressure at the bottom of the fluid is greater than that at the top. We can also state that the pressure exerted on a small object in the fluid is the same regardless of the orientation of the object.

As an example of Pascal's principle, consider the hydraulic press shown in Figure 9.1. The small-area piston A_1 has an external force F_1 applied to it. At the other end, the large-area piston A_2 has some unknown force F_2 acting on it. How do these forces compare? According to Pascal's principle, the force per unit area represents an external pressure that will be transmitted uniformly through the fluid. Thus, we can write

$$\frac{F_1}{A_1} = \frac{F_2}{A_2}$$

Sample Problem

Referring to Figure 9.1, suppose a force of 10 N is applied to the small piston of area 0.05 m². If the large piston has an area of 0.15 m², what is the maximum weight the large piston can lift?

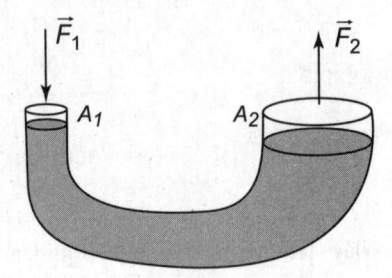

Figure 9.1

Solution

Since the secondary force is proportional to the ratio of the areas, $F_2 = 30$ N.

Recall that density (ρ) is defined by

$$\rho = \frac{\text{mass}}{\text{volume}}$$

Static Pressure and Depth

Figure 9.2 shows a tall column of liquid in a sealed container. What is the pressure exerted on the bottom of the container? To answer this question, we first consider the weight of the column of liquid of height h. Since $F_g = mg$ and $m = \rho V$, the weight is $\rho g V$. This is the force applied to the bottom of the container.

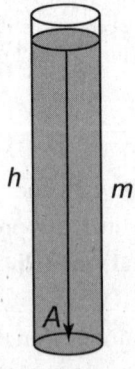

Figure 9.2 Column of Fluid

Now, in a container with a regular shape, $V = Ah$, where A is the cross-sectional area (in this case, we have a cylinder whose cross-sections are uniform circles). Thus, $F = \rho g V = \rho g A h$. Using the definition of pressure, we obtain

$$P = \frac{F}{A} = \rho g h$$

If the container is open at the top, then air pressure adds to the pressure of the column of liquid. The total pressure can therefore be written as $P = P_{ext} + \rho g h$. Note that the pressure is a function of depth only, not container width or size.

▶ Sample Problem

A column of mercury is held up at 1 atm of pressure in an open-tube barometer (see the accompanying diagram). To what height does it rise? The density of mercury is 13.6 times the density of water.

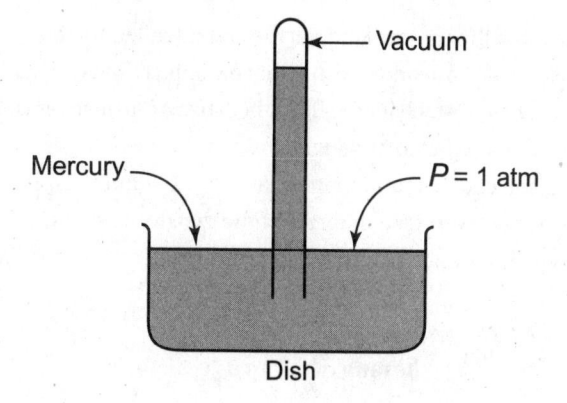

✓ Solution

At 1 atm the pressure is 101 kPa. Thus, we can write

$$1.01 \times 10^5 \text{ N/m}^2 = (13.6 \times 10^3 \text{ kg/m}^3)(9.8 \text{ m/s}^2) h$$

$$h = 0.76 \text{ m} = 76 \text{ cm}$$

Buoyancy and Archimedes' Principle

When an object is immersed in water, it experiences an upward force. In a cylinder filled with water, the action of inserting a mass in the liquid causes some of the liquid to displace upward. The volume of the water displaced is equal to the volume of the object (even if it is irregularly shaped), as illustrated in Figure 9.3.

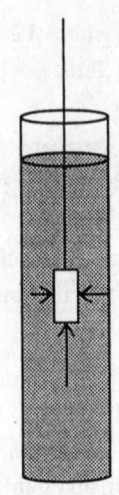

Figure 9.3 Measuring Volume by Water Displacement

Archimedes' principle states that the upward force of the water (called the **buoyant force**), F_B, is equal to the weight of the water displaced. Normally, one might think that an object floats if its density is less than that of water. This statement is only partially correct. A steel needle floats because of surface tension, and a steel ship floats because it displaces a volume of water equal to its weight.

The weight of the water displaced can be found mathematically. The fluid displaced has a weight $F_g = mg$. Now, the mass can be expressed in terms of the density of the liquid and its volume, $m = \rho V$. Hence, $F_g = \rho V g$. By letting V_d represent the volume of displaced water, we obtain:

$$F_B = F_g = \rho g V_d$$

> **TIP**
> A submerged object displaces a volume of water equal to its own volume.

The volume of the object can be determined in terms of the apparent loss of weight in water. Suppose an object weighs 5 N in the air and 4.5 N when submerged in water. The difference of 0.5 N is the weight of the water displaced. The volume is therefore given by

$$V_f = \frac{\Delta m}{\rho_{fluid}} = \frac{\Delta F_g}{g \rho}$$

Using this relationship, we have $V_f = (0.5 \text{ N})/(9.8 \text{ N/kg})(1 \times 10^3 \text{ kg/m}^3) = 5.1 \times 10^{-5} \text{ m}^3$.

Specific gravity or relative density is a useful tool for comparing one fluid to another. It is the ratio of the density of the fluid in question to a reference fluid. For liquids, the reference is usually water. For gases, the reference is usually air. If a liquid has a relative density less than 1.0 it will float in water. For example, the specific gravity of ice is 0.917, which means that ice floats since it need not displace as much water in order to get an adequate buoyant force. This value of relative density also gives the fraction of ice that will be submerged: 91.7% (or, put another way, 8.3% of the ice is above the water line in a body of fresh water).

Fluids in Motion

The situation regarding static pressures in fluids changes when they are in motion. Microscopically, we could try to account for the motion of all molecular particles that make up the fluid, but this would not be very practical. Instead, we treat the fluid as a whole and consider what happens as the fluid passes through a given cross-sectional area each second. Sometimes the word *flux* is used to describe the volume of fluid passing through a given area each second.

Consider the fluid shown in Figure 9.4 moving uniformly with a velocity v in a time t through a segment of a cylindrical pipe. The distance traveled is given by the product vt. Since the motion is ideally smooth, there is no resistance offered by the fluid as different layers move relative to one another. This resistance is known as **viscosity**, and the type of fluid motion we are considering here is called **laminar flow**.

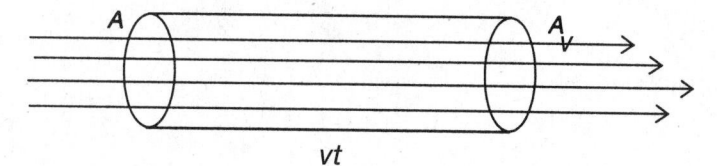

Figure 9.4 Laminar Flow

The rate of flow Q is defined to be the volume of fluid flowing out of the pipe each second (in m³/s):

$$Q = \frac{(\text{Volume})}{t} = \frac{vtA}{t} = vA$$

If the flow is laminar, then the **equation of continuity** states that the rate of flow Q will remain constant. Therefore, as the cross-sectional area decreases, the velocity must increase:

$$Q_1 = Q_2$$
$$v_1 A_1 = v_2 A_2$$

Bernoulli's Equation

Consider a fluid moving through an irregularly shaped tube at two different levels given by h_1 and h_2 as shown in Figure 9.5. At the lower level, the fluid exerts a pressure P_1 while moving through an area A_1 with a velocity v_1. At the top, the fluid exerts a pressure P_2 while moving through an area A_2 with a velocity v_2. Bernoulli's equation is related to changes in pressure as a function of velocity.

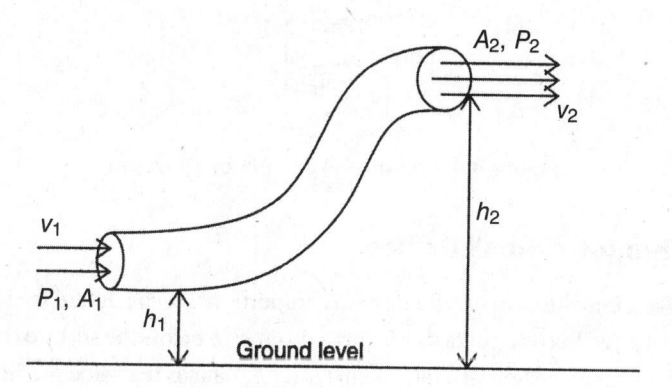

Figure 9.5 Bernoulli Variables

Let us begin by considering the work done in moving the fluid from position 1 to position 2. The power generated is equal to the product of the pressure and the rate of flow (PR), and so the work, which is equal to the product of power and time, can be written as

$$W = P_1 A_1 v_1 t - P_2 A_2 v_2 t$$

The change in the potential energy is given by

$$\Delta PE = mgh_2 - mgh_1$$

The change in kinetic energy is given by

$$\Delta KE = \tfrac{1}{2} m v_2^2 - \tfrac{1}{2} m v_1^2$$

Adding up both changes in energy and equating it with the work done, we obtain

$$W = mgh_2 - mgh_1 - \tfrac{1}{2} m v_2^2 - \tfrac{1}{2} m v_1^2$$

The work done is also equal to

$$W = (P_2 - P_1)V$$

Setting these two expressions for work equal to each other and simplifying gives us

$$P_1 + \rho g h_1 + \tfrac{1}{2}\rho v_1^2 = P_2 + \rho g h_2 + \tfrac{1}{2}\rho v_2^2$$

The last equation is known as **Bernoulli's equation**. Now, let us consider some applications of this equation.

> **TIP**
> Compare this equation and concept to the equation for the conservation of mechanical energy.

A Fluid at Rest

In Figure 9.6, we see a static fluid. The two layers at heights h_1 and h_2 have static pressures P_1 and P_2. Since the fluid is at rest ($v_1 = v_2 = 0$), Bernoulli's equation reduces to

$$\Delta P = P_2 - P_1 = \rho g \Delta h = \rho g (h_2 - h_1)$$

The difference in pressure is just proportional to the difference in levels.

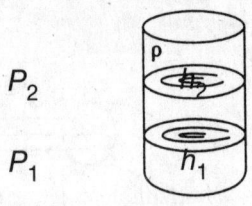

Figure 9.6 Pressure as a Function of Depth

A Fluid Escaping Through a Small Orifice

Earlier in this chapter, we saw that the pressure difference is proportional to the difference in height. Suppose a small hole of circular area A is punched into the container below a distance h below the surface (Figure 9.7). This pressure difference will force the fluid out of the hole at a rate of flow $R = vA$. What is the velocity of the fluid as it escapes? And what is the rate of flow?

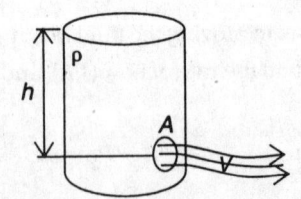

Figure 9.7 Fluid Flow from Pressure

To answer these questions, we consider Bernoulli's equation as an analog of the conservation of energy equation. Let's choose the potential energy to be zero at the hole (i.e., set $h = 0$ at this height). The top of the fluid is essentially at rest. In addition, the pressure at the top and the pressure at the orifice are both the same: atmospheric pressure. Therefore, the pressure term cancels. Thus, using Bernoulli's equation, we can solve for the flow rate:

$$\frac{1}{2}\rho v^2 = \rho g h$$
$$v = \sqrt{2gh}$$
$$R = vA = A\sqrt{2gh}$$

A Fluid Moving Horizontally

Consider a fluid moving horizontally through a tube that narrows in area. This is known as the Venturi effect. Bernoulli's equation states that as the velocity of a moving fluid increases, its static pressure decreases. We can analyze this in Figure 9.8.

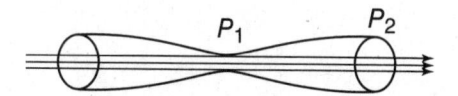

Figure 9.8 Venturi Tube

Since the level is horizontal, $h_1 = h_2$, so we can eliminate the $\rho g h$ term. We can therefore write

$$P_1 + \frac{1}{2}\rho v_1^2 = P_2 + \frac{1}{2}\rho v_2^2$$
$$\Delta P = P_1 - P_2 = \frac{1}{2}\rho(v_2^2 - v_1^2)$$

Sample Problem

Water ($\rho = 1{,}000$ kg/m^3) is flowing smoothly through a horizontal pipe that tapers from 1.5×10^{-3} m^2 to 0.8×10^{-3} m^2 in a cross-sectional area. The pressure difference between the two sections is equal to 5,000 Pa. What is the volume flow rate of the water?

Solution

Since the water flows smoothly, we know that the volume flow rate is constant:

$$R = v_1 A_1 = v_2 A_2$$

We also know that from Bernoulli's equation

$$P_1 + \frac{1}{2}\rho v_1^2 + \rho g y_1 = P_2 + \frac{1}{2}\rho v_2^2 + \rho g y^2$$

But, since the pipe is horizontal, $y_1 = y_2$, and so the equation simplifies to

$$P_1 + \frac{1}{2}\rho v_1^2 = P_2 + \frac{1}{2}\rho v_2^2$$

We can simplify this by eliminating the velocities in each expression since

$$v_1 = \frac{R}{A_1} \quad \text{and} \quad v_2 = \frac{R}{A_2}$$

Thus,
$$P_1 + \frac{1}{2}\rho\frac{R^2}{A_1^2} = P_2 + \frac{1}{2}\rho\frac{R^2}{A_2^2}$$

We also know that $P_1 > P_2$ since pressure decreases with increasing velocity, and velocity increases with decreasing area. Thus, we can write:
$$(P_1 - P_2) = \frac{1}{2}\rho R^2\left(\frac{1}{A_2^2} - \frac{1}{A_1^2}\right)$$

Since we know all values (including the pressure difference $P_1 - P_2$), we can substitute and solve for R, and we obtain $R = 2.99 \times 10^{-3}$ m³/s.

In aerodynamics, a wing moving in level flight has a lifting force acting on it exactly equal to its load. This force is caused partially by the pressure difference between the upper and lower surfaces of the wing. For a variety of different reasons, the airflow above the wing is faster than the airflow below the wing (see Figure 9.9). By Bernoulli's equation, this faster airflow corresponds to a lower air pressure on the top of the wing. The difference between the higher pressure on the bottom and the lower pressure on the top is a contributing factor to the lift force on the wing.

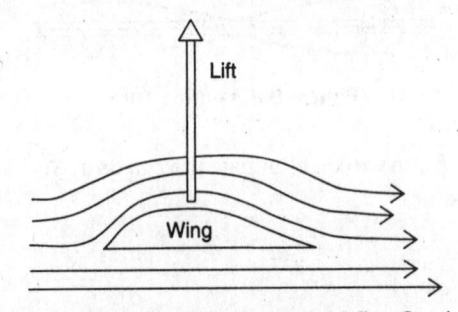

Figure 9.9 Airflow Pattern over a Wing Section

SUMMARY

- Liquids are called incompressible fluids, whereas gases are called compressible fluids.
- Pascal's principle states that in a confined fluid at rest, any change in pressure is transmitted undiminished throughout the fluid.
- The pressure in a confined fluid is proportional to the density of the fluid and its depth ($P = \rho gh$). A fluid open to the air has the pressure on top as well (P_0):
$$P = P_0 + \rho gh$$
- Archimedes' principle states that a submerged object will displace a volume of water equal to its own volume. A submerged object also experiences an upward force called the buoyant force, which is equal to the weight of the water displaced.
- An object will neither rise nor sink if it displaces a volume of water equal in weight to its own weight in the air.
- For laminar flow, $A_1v_1 = A_2v_2$.
- Bernoulli's principle states that for a fluid in motion, the static pressure will decrease with an increase in velocity. This principle helps explain the lifting force of an airplane wing. Bernoulli's principle is energy conservation for fluids: $P + \rho gh + \frac{1}{2}\rho v^2 =$ constant throughout the fluid.

Problem-Solving Strategies for Fluids

Solving fluid problems is similar to solving particle problems. We treat the fluid as a whole unit (i.e., macroscopically) as opposed to microscopically. The units of pressure must be either pascals or N/m^2 in order to use the formulas derived.

The concepts of Bernoulli's principle and Archimedes' principle should be thoroughly understood, as well as their applications and implications. Buoyancy is an important physical phenomenon and an important part of your overall physics education.

As always, drawing a sketch helps. Often, conceptual knowledge will be enhanced if you understand how the variables in a formula are related. Consider questions that involve changing one variable and observing the effect on others.

Practice Exercises

Multiple-Choice

1. The rate of flow of a liquid from a hole in a container depends on all of the following except

   (A) the density of the liquid
   (B) the height of the liquid above the hole
   (C) the area of the hole
   (D) the acceleration of gravity

2. A person is standing on a railroad station platform when a high-speed train passes by. The person will tend to be

   (A) pushed away from the train
   (B) pulled in toward the train
   (C) pushed upward into the air
   (D) pulled down into the ground

3. Bernoulli's equation is based on which law of physics?

   (A) Conservation of linear momentum
   (B) Conservation of angular momentum
   (C) Newton's first law of motion
   (D) Conservation of energy

4. Which of the following expressions represents the power generated by a liquid flowing out of a hole of area A with a velocity v?

   (A) (pressure) × (rate of flow)
   (B) (pressure)/(rate of flow)
   (C) (rate of flow)/(pressure)
   (D) (pressure) × (velocity)/(area)

5. Which of the following statements is an expression of the equation of continuity?

   (A) Rate of flow equals the product of velocity and cross-sectional area.
   (B) Rate of flow depends on the height of the fluid above the hole.
   (C) Pressure in a static fluid is transmitted uniformly throughout.
   (D) Fluid flows faster through a narrower pipe.

6. A moving fluid has an average pressure of 600 Pa as it exits a circular hole with a radius of 2 cm at a velocity of 60 m/s. What is the approximate power generated by the fluid?

   (A) 32 W
   (B) 62 W
   (C) 1,200 W
   (D) 45 W

7. Alcohol has a specific gravity of 0.79. If a barometer consisting of an open-ended tube placed in a dish of alcohol is used at sea level, to what height in the tube will the alcohol rise?

   (A) 8.1 m
   (B) 7.9 m
   (C) 15.2 m
   (D) 13.1 m

8. An ice cube is dropped into a mixed drink containing alcohol and water. The ice cube sinks to the bottom. From this, you can conclude

   (A) that the drink is mostly alcohol
   (B) that the drink is mostly water
   (C) that the drink is equally mixed with water and alcohol
   (D) nothing, unless you know how much liquid is present

9. A 2-N force is used to push a small piston 10 cm downward in a simple hydraulic machine. If the opposite large piston rises by 0.5 cm, what is the maximum weight the large piston can lift?

   (A) 2 N
   (B) 40 N
   (C) 20 N
   (D) 4 N

10. Balsa wood with an average density of 130 kg/m³ is floating in pure water. What percentage of the wood is submerged?

   (A) 87%
   (B) 13%
   (C) 50%
   (D) 25%

Additional Practice

1. A U-tube open at both ends is partially filled with water. Benzene ($\rho = 0.897 \times 10^3$ kg/m³) is poured into one arm, forming a column 4 cm high. What is the difference in height between the two surfaces?

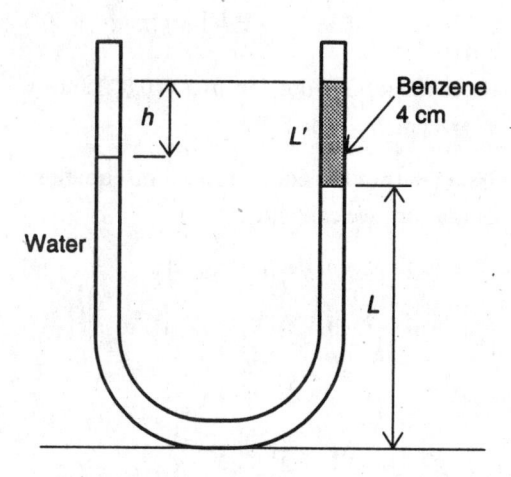

2. A Venturi tube has a pressure difference of 15,000 Pa. The entrance radius is 3 cm, while the exit radius is 1 cm. What are the entrance velocity, exit velocity, and flow rate if the fluid is gasoline ($\rho = 700$ kg/m³)?

3. A cylindrical tank of water (height H) is punctured at a height h above the bottom. How far from the base of the tank will the water stream land (in terms of h and H)? What must the value of h be such that the distance at which the stream lands will be equal to H?

4. Which has more pressure on the bottom—a large tank of water 30 cm deep or a cup of water 35 cm deep? Explain your answer.

5. Two paper cups are suspended by strings and hung near each other. They are separated by about 10 cm. When you blow air between them, the cups are attracted to one another. Explain why this occurs.

Answers Explained

Multiple-Choice Problems

1. **(A)** The flow rate of a liquid from a hole does not depend on the density of the liquid.

2. **(B)** Because of the Bernoulli effect, the speeding train reduces the air pressure between the person and the train. This pressure difference creates a force tending to pull the person into the train. Be very careful when standing on a train platform!

3. **(D)** Bernoulli's principle was developed as an application of conservation of energy.

4. **(A)** Power is expressed in J/s. The product of pressure (force divided by area) and flow rate (m^3/s) leads to newtons times meters over seconds (J/s).

5. **(D)** The equation of continuity states that the product of velocity and area is a constant for a given fluid. A consequence of this is that a fluid moves faster through a narrower pipe.

6. **(D)** The power is equal to the product of the pressure and the flow rate. The flow rate is equal to the product of the velocity and the cross-sectional area (which is a circle of radius 0.02 m). The area is given by $\pi r^2 = 1.256 \times 10^{-3}\ m^2$. When we multiply this area by the velocity and the pressure, we get 45.2 W as a measure of the generated power.

7. **(D)** The pressure exerted by the air is balanced by the column of liquid alcohol in equilibrium:
$$h = \frac{P}{\rho g} = \frac{1.01 \times 10^5\ N/m^2}{(790\ kg/m^3)(9.8\ m/s^2)} = 13.1\ m$$

8. **(A)** Since the density of alcohol is less than that of water, ice floats "lower" in alcohol than in water. From the given information, the drink appears to be mostly alcohol.

9. **(B)** We need to use the conservation of work-energy in this problem. The work done on the small piston must equal the work done by the large piston. Since the ratio of displacements is 20:1, the large piston will be able to support a maximum load of 40 N (since 2 N × 20 = 40 N).

10. **(B)** The percentage submerged is given by the ratio of its density to that of pure water ($1{,}000\ kg/m^3$). Thus, $130/1{,}000 = 0.13 = 13\%$.

Additional Practice

1. Let L' be the level of benzene that will float on top of the water, let L be the level of water, and let h be the difference in levels. Since the tube is open, the pressures are equalized at both ends. Thus, we can write
$$(L + L' - h)g\rho_w = L'g\rho_b + Lg\rho_w$$
$$h = L'\left(1 - \frac{\rho_b}{\rho_w}\right)$$

Since the ratio of the densities is 0.879 and $L' = 4$ cm, $h = 0.484$ cm.

2. Using Bernoulli's equation and the equation of continuity, we can write
$$\Delta P = \frac{\rho}{2}\left(v_2^2 - v_1^2\right)$$
$$v_1 A_1 = v_2 A_2$$
$$v_2 = v_1 \frac{A_1}{A_2}$$
$$\Delta P = \frac{\rho}{2} v_1^2 \left(1 - \frac{A_1^2}{A_2^2}\right)$$

From the given information, we know that the ratio of areas $A_1/A_2 = 9$. Thus, substituting all given values into the last equation yields $v_1 = 0.732$ m/s, $R = 0.0021\ m^3/s$, and $v_2 = 6.56$ m/s.

3. The change in potential energy must be equal to the change in horizontal kinetic energy:
$$mg(H - h) = \left(\frac{1}{2}\right)m v_x^2$$
$$v_x = \sqrt{2g(H - h)}$$

Now if we assume, as in projectile motion, that the horizontal velocity remains the same and the only acceleration of the stream is vertically downward because of gravity, we can write

$$x = v_x t$$
$$y = h = \tfrac{1}{2} g t^2$$
$$t = \sqrt{\tfrac{2h}{g}}$$
$$x = 2\sqrt{h(H - h)}$$

For $x = H$, we must have $h = H/2$, as can be easily verified using the preceding range formula.

4. The pressure at the bottom of a container of water depends on its depth and not on its volume. Thus, the pressure at the bottom of the 35-cm cup is greater.

5. Bernoulli's principle states that as the velocity of a moving fluid increases, the pressure it exerts decreases. Thus, blowing between the cups reduces the air pressure between them, causing a net force that pushes them together.

Practice Tests

ANSWER SHEET
Practice Test 1

1. Ⓐ Ⓑ Ⓒ Ⓓ
2. Ⓐ Ⓑ Ⓒ Ⓓ
3. Ⓐ Ⓑ Ⓒ Ⓓ
4. Ⓐ Ⓑ Ⓒ Ⓓ
5. Ⓐ Ⓑ Ⓒ Ⓓ
6. Ⓐ Ⓑ Ⓒ Ⓓ
7. Ⓐ Ⓑ Ⓒ Ⓓ
8. Ⓐ Ⓑ Ⓒ Ⓓ
9. Ⓐ Ⓑ Ⓒ Ⓓ
10. Ⓐ Ⓑ Ⓒ Ⓓ
11. Ⓐ Ⓑ Ⓒ Ⓓ
12. Ⓐ Ⓑ Ⓒ Ⓓ
13. Ⓐ Ⓑ Ⓒ Ⓓ
14. Ⓐ Ⓑ Ⓒ Ⓓ
15. Ⓐ Ⓑ Ⓒ Ⓓ
16. Ⓐ Ⓑ Ⓒ Ⓓ
17. Ⓐ Ⓑ Ⓒ Ⓓ
18. Ⓐ Ⓑ Ⓒ Ⓓ
19. Ⓐ Ⓑ Ⓒ Ⓓ
20. Ⓐ Ⓑ Ⓒ Ⓓ
21. Ⓐ Ⓑ Ⓒ Ⓓ
22. Ⓐ Ⓑ Ⓒ Ⓓ
23. Ⓐ Ⓑ Ⓒ Ⓓ
24. Ⓐ Ⓑ Ⓒ Ⓓ
25. Ⓐ Ⓑ Ⓒ Ⓓ
26. Ⓐ Ⓑ Ⓒ Ⓓ
27. Ⓐ Ⓑ Ⓒ Ⓓ
28. Ⓐ Ⓑ Ⓒ Ⓓ
29. Ⓐ Ⓑ Ⓒ Ⓓ
30. Ⓐ Ⓑ Ⓒ Ⓓ
31. Ⓐ Ⓑ Ⓒ Ⓓ
32. Ⓐ Ⓑ Ⓒ Ⓓ
33. Ⓐ Ⓑ Ⓒ Ⓓ
34. Ⓐ Ⓑ Ⓒ Ⓓ
35. Ⓐ Ⓑ Ⓒ Ⓓ
36. Ⓐ Ⓑ Ⓒ Ⓓ
37. Ⓐ Ⓑ Ⓒ Ⓓ
38. Ⓐ Ⓑ Ⓒ Ⓓ
39. Ⓐ Ⓑ Ⓒ Ⓓ
40. Ⓐ Ⓑ Ⓒ Ⓓ

Practice Test 1

Section I: Multiple-Choice

TIME: 80 MINUTES
40 QUESTIONS

DIRECTIONS: Each of the questions or incomplete statements below is followed by four suggested answers or completions. Select the one that is best in each case. You have 80 minutes to complete this portion of the test. You may use a calculator and the information sheets provided in the appendix.

1. Two objects are thrown vertically upward from the same initial height. One object has twice the initial velocity of the other. Neglecting any air resistance, the object with the greater initial velocity will rise to a maximum height that is

   (A) twice that of the other object, assuming they have the same mass
   (B) twice that of the other object, independent of their masses
   (C) four times that of the other object, assuming they have the same mass
   (D) four times that of the other object, independent of their masses

2. A 2-kilogram cart has a velocity of 4 meters per second to the right. It collides with a 5-kilogram cart moving to the left at 1 meter per second. After the collision, the two carts stick together. Can the magnitude and the direction of the velocity of the two carts after the collision be determined from the given information?

   (A) No, since the collision is inelastic, we must know the energy lost.
   (B) Yes, the collision is elastic: 3/7 m/s left.
   (C) Yes, the collision is inelastic: 3/7 m/s right.
   (D) Yes, the speed is not 3/7 m/s.

3. Projectile X is launched at a 30-degree angle above the horizon with a speed of 100 m/s. Projectile Y is launched at a 60-degree angle with the same speed. Which of the following correctly compares the horizontal range and maximum altitude obtained by these two projectiles?

	Range	Altitude
(A)	X goes farther	X goes higher
(B)	Y goes farther	Y goes higher
(C)	X goes farther	Y goes higher
(D)	X and Y equal	Y goes higher

4. A 5-kg mass is sitting at rest on a horizontal surface. A horizontal force of 10 N will start the mass moving. What is the best statement about the coefficient and type of friction between the mass and the surface?

   (A) >0.20 static
   (B) <0.20 static
   (C) <0.20 kinetic
   (D) >0.20 kinetic

5. Which of the following graphs represents an object *moving* with no net force acting on it?

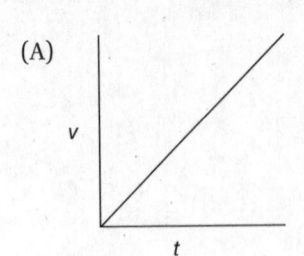

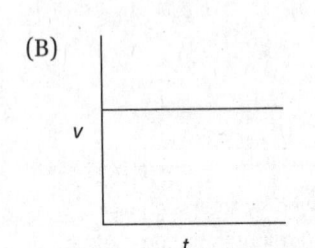

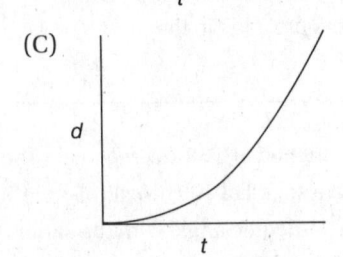

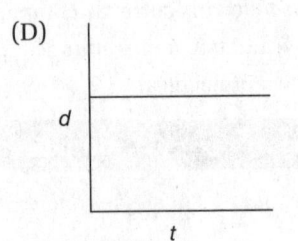

(A) [graph of v vs t, linearly increasing from origin]

(B) [graph of v vs t, constant positive value]

(C) [graph of d vs t, curving upward]

(D) [graph of d vs t, constant positive value]

6. A projectile is launched horizontally with an initial velocity v_0 from a height h. If it is assumed that there is no air resistance, which of the following expressions represents the vertical position of the projectile? In other words, $y(x) = ?$

(A) $h - gv_0^2/2x^2$
(B) $h - gv/2v_0^2$
(C) $h - gx^2/2v_0^2$
(D) $h - gx^2/v_0^2$

Questions 7 and 8 are based on the information and diagram below:

A 0.4-kilogram mass is oscillating on a spring that has a force constant of $k = 1{,}000$ newtons per meter.

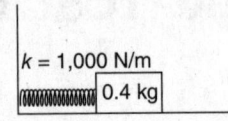

[diagram: spring with $k = 1{,}000$ N/m attached to 0.4 kg mass]

7. Which of the following measurements would allow you to determine the maximum velocity experienced by the mass?

(A) No additional information is required.
(B) Minimum velocity
(C) Maximum acceleration
(D) None of these would allow you to determine maximum velocity.

8. Which of the following statements concerning the oscillatory motion described above is correct? (All statements refer to magnitudes.)

(A) The maximum velocity and maximum acceleration occur at the same time.
(B) The maximum velocity occurs when the acceleration is a minimum.
(C) The velocity is always directly proportional to the displacement.
(D) The maximum velocity occurs when the displacement is a maximum.

9. A bullet of known mass (m_1) is fired vertically into an initially stationary wood block of known mass (m_2). The resulting wood + bullet combined system is then measured to rise to a maximum height of h. Can the initial speed of the bullet be calculated from this information?

(A) Yes. Solve $(m_1 + m_2)gh = \frac{1}{2}(m_1)v^2$.
(B) Yes. Solve the momentum conservation of collision first and the energy conservation of the rising combination second.
(C) No. We don't know if momentum is conserved during this collision.
(D) No. We don't know enough details about the mechanical energy lost during the collision.

10. A 10-kg mass is being pushed horizontally by a constant force along a rough surface ($\mu_k = 0.1$) at constant velocity. Which of the following is the best statement regarding the constant force (f)?

    (A) $f = 10$ N
    (B) $f > 10$ N
    (C) $f < 10$ N
    (D) $f > 1$ N

11. A 10-newton force is applied to two masses, 4 kilograms and 1 kilogram, respectively, that are in contact as shown below. The horizontal motion is along a frictionless plane. What is the magnitude of the contact force between the two masses?

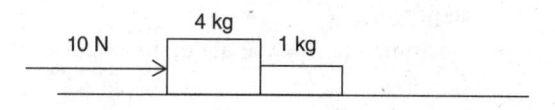

    (A) 10 N
    (B) 8 N
    (C) 6 N
    (D) 2 N

12. An object with mass m is dropped from height h above the ground. While neglecting air resistance, which formula best describes the power generated if the object takes time t to fall?

    (A) mgh
    (B) $mght$
    (C) $mg^2t/2$
    (D) mgh/t

13. A 1,500-kilogram car has a velocity of 25 meters per second. If it is brought to a stop by a nonconstant force in 10 to 15 seconds, can the magnitude of the impulse applied be determined?

    (A) Yes, it is 37,500 N · s.
    (B) No, you need to know the details about the nonconstant force.
    (C) No, you must know the exact duration of the impulse.
    (D) No, you must know the average force during and the duration of the impulse.

14. A block of mass M rests on a rough incline, as shown below. The angle of elevation of the incline is increased until an angle of θ is reached. At that angle, the mass begins to slide down the incline. Which of the following is an expression for the coefficient of static friction μ?

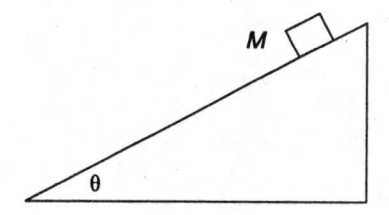

    (A) $\tan \theta$
    (B) $\sin \theta$
    (C) $\cos \theta$
    (D) $1/(\cos \theta)$

15. A pendulum of a given length swings back and forth a certain number of times per second. If the pendulum now swings back and forth the same number of times but in twice the time, the length of the pendulum should be

    (A) doubled
    (B) quartered
    (C) quadrupled
    (D) halved

182　AP PHYSICS 1

16. This graph of force versus time shows how the force acts on an object of mass m for a total time of T seconds. If the mass begins at rest, which is the correct method to find the final speed of the mass?

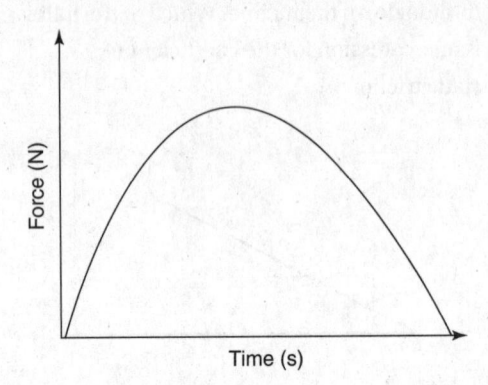

    (A) Average value of this graph times total time divided by mass
    (B) Area under this graph divided by mass
    (C) Since the final force is zero, the object is at rest after time T
    (D) Average slope of this graph divided by mass

17. A child of unknown mass is on a swing of unknown length that varies in height from 75 cm at its lowest height above the ground to a maximum height of 225 cm above the ground. Is there enough information to find the speed of the swing at its lowest point?

    (A) No, the child's mass must be known.
    (B) No, the length of the swing must be known to determine the centripetal acceleration.
    (C) Yes, it is 5.5 m/s.
    (D) Yes, it is 4 m/s.

18. The gravitational force of attraction between two identical masses is 36 N when the masses are separated by a distance of 3 m. If the distance between them is reduced to 1 m, which of the following is true about the net gravitational field strength due to both masses being at the halfway point?

    (A) It is 9 times stronger total.
    (B) Not enough information is given to determine net gravitational field strength.
    (C) Each mass's gravitation is 9 times stronger, so the net gravitational field strength is 18 times stronger.
    (D) It is zero.

19. A 200-kilogram cart rests on top of a frictionless hill as shown below. Can the impulse required to stop the cart when it is at the top of the 10-meter hill be calculated?

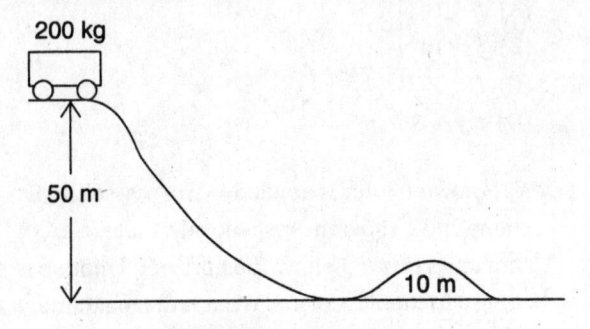

    (A) No, more information about friction is required.
    (B) No, more information about the impulse is required.
    (C) Yes, calculate the velocity from free-fall kinematics and use this velocity in the change in momentum equation.
    (D) Yes, calculate the speed from energy conservation and use this speed in the change in momentum equation.

20. An object of mass m starts at a height of H_1 with a speed of v_1. A few minutes later, it is at a height of H_2 and a speed v_2. Which of the following expressions best represents the work done to the mass by nongravitational forces to the object during this time?

    (A) $mg(H_2 - H_1) + \frac{1}{2}m(v_2^2 - v_1^2)$
    (B) $mg(H_2 - H_1) - \frac{1}{2}m(v_2^2 - v_1^2)$
    (C) $\frac{1}{2}m(v_2^2 - v_1^2)$
    (D) $\frac{1}{2}m(v_1^2 - v_2^2)$

21. An object with 0.2-kg mass is pushed down vertically onto an elastic spring ($k = 20$ N/m). The spring is compressed by 20 cm and then released such that the object will fly off. Which of the following will have the largest effect on increasing the maximum height the object will fly? (Assume no air resistance.)

    (A) Halving the mass
    (B) Doubling the compression distance
    (C) Using a spring with a spring constant twice as big
    (D) Doing the same experiment on a different planet with half the gravitational field strength

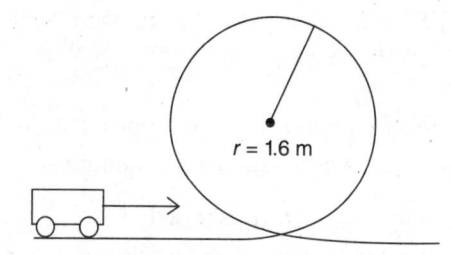

22. A cart with a mass of m needs to complete a loop-the-loop of radius r, as shown above. What is the approximate minimum velocity required to achieve this goal?

    (A) $(gr)^{1/2}$
    (B) $(5gr)^{1/2}$
    (C) $2(gr)^{1/2}$
    (D) $(2gr)^{1/2}$

23. A particle traveling in the direction shown below experiences the force shown. In this situation, the particle will

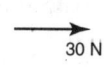

    (A) slow down only
    (B) speed up only
    (C) turn and speed up
    (D) neither turn nor change speed

24. Which of the following velocity component versus time graphs match a person riding an elevator steadily downward on a cruise boat that is accelerating out of port (assume y is vertical and x is horizontal)?

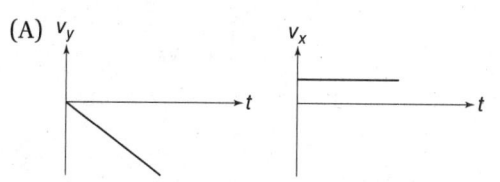

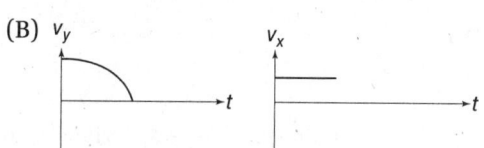

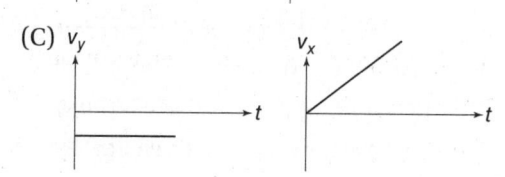

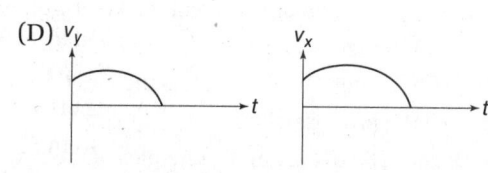

    (A) v_y , v_x
    (B) v_y , v_x
    (C) v_y , v_x
    (D) v_y , v_x

25. If a mass of 9 kg begins at rest and experiences the following net force, what is its speed after 3 seconds?

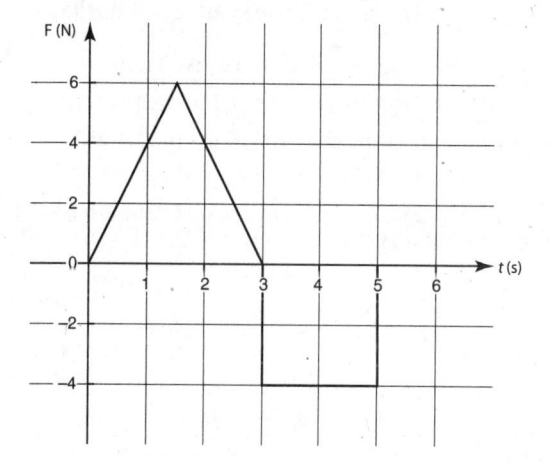

    (A) 0 m/s
    (B) 1 m/s
    (C) 2 m/s
    (D) 4 m/s

26. Two blocks are sliding to the right as shown below. If they are brought to a stop by a force F applied from the right, what is the net work done to the two blocks as they are brought to a stop?

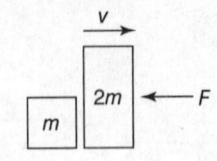

   (A) $-1.5\ mv^2$
   (B) $-0.5\ mv^2$
   (C) $-Fv$
   (D) $-mv^2$

27. A satellite is launched to twice the orbital radius of a similar satellite (both are in stable circular Earth orbits). Compare the kinetic energies of the two satellites.

   (A) They have the same kinetic energy.
   (B) The lower satellite has half the kinetic energy of the higher satellite.
   (C) The higher satellite has half the kinetic energy of the lower satellite.
   (D) The ratio of the kinetic energies of the two satellites is a factor of four.

28. It has been documented that a major earthquake on Earth can change the time it takes the planet to rotate fully. From this, one can conclude that

   (A) the Earth is not an isolated system
   (B) a net torque was applied to the planet
   (C) the rotational inertia of the planet was changed
   (D) work was done to the planet somehow

29. Compare the two columns of water pictured below. They have the same height. Which of the following statements compares the total weight and the pressure at the bottom of the columns?

   (A) Weight and pressure are greater for the wider column.
   (B) Weight and pressure are the same for both.
   (C) Weight is greater for the wider column, while pressure is the same.
   (D) Weight is the same, while pressure is greater for the wider column.

30. An otherwise similar boat will ride higher in a fluid (i.e., have less submerged volume) if

   (A) the fluid's density is raised
   (B) the fluid's temperature is raised
   (C) the boat's density is raised
   (D) the boat shifts its cargo to a lower deck

31. Fluid is poured into an open U-shaped tube as shown below. If a person blows across the top of the right-hand side's opening, what will happen to the fluid in the tube?

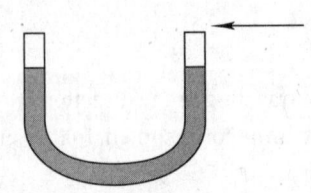

   (A) It will remain the same.
   (B) It will rise on the right-hand side and lower on the left-hand side.
   (C) It will rise on both sides.
   (D) It will rise on the left-hand side and lower on the right-hand side.

32. A boat moving due north crosses a river 240 meters wide with a velocity of 8 meters per second relative to the water. The river flows east with a velocity of 6 meters per second. How far downstream will the boat be when it has crossed the river?

    (A) 240 m
    (B) 180 m
    (C) 420 m
    (D) 300 m

33. A variable force acts on a 2-kilogram mass according to the graph below.

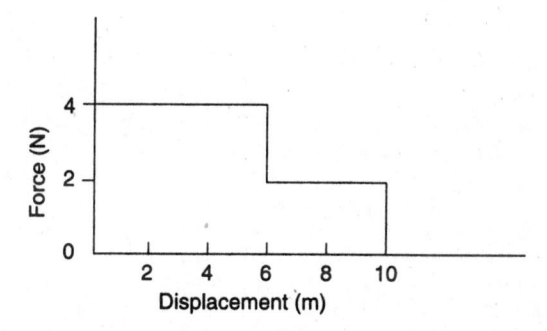

    How much work was done while displacing the mass 10 meters?

    (A) 40 J
    (B) 38 J
    (C) 32 J
    (D) 30 J

34. A man weighing himself is standing on a bathroom scale in an elevator that is accelerating upward at a rate of 0.5 meter per second squared. By what percentage is the reading of the scale off from the person's true weight?

    (A) 0% (accurate)
    (B) 5% too high
    (C) 5% too low
    (D) 0.5% too high

35. A 50-kilogram person is sitting on a seesaw 1.2 meters from the balance point. On the other side, a 70-kilogram person is balanced. How far from the balance point is the second person sitting?

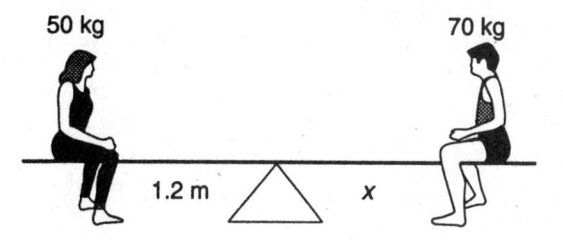

    (A) 0.57 m
    (B) 0.75 m
    (C) 0.63 m
    (D) 0.86 m

36. An object rolls down a steep incline with very little friction. At the same time, an object of equal mass slides down a similar but frictionless incline. Which one takes less time to get to the bottom?

    (A) They take the same time.
    (B) The rolling object takes less time.
    (C) The sliding object takes less time.
    (D) The answer depends on the rolling object's rotational inertia.

37. What is the value of g at a position above Earth's surface equal to Earth's radius?

    (A) 9.8 N/kg
    (B) 4.9 N/kg
    (C) 2.45 N/kg
    (D) 1.6 N/kg

38. If an object is spinning at 150 RPM (revolutions per minute) and comes to a stop in 2 seconds, what is its average acceleration in radians/s^2?

    (A) -2.5π
    (B) -2π
    (C) -5π
    (D) -10π

39. What is the work done by a horizontal spring (spring constant k) expanding from a compression distance x to an extension distance x to an attached mass?

(A) $2kx^2$
(B) $\frac{1}{2}kx^2$
(C) kx^2
(D) 0

40. If an isolated spinning object's rotational inertia is reduced by a factor of 3 by internal forces, how will its angular momentum change?

(A) Angular momentum will be 3 times its previous value.
(B) Angular momentum will be reduced to 1/3 its previous value.
(C) Angular momentum will be reduced to 1/9 its previous value.
(D) Angular momentum will remain unchanged.

Section II: Free-Response

TIME: 100 MINUTES
4 QUESTIONS

DIRECTIONS: You have 100 minutes to complete this portion of the test. You may use a calculator and the information sheets provided in the appendix.

1. 10 points; suggested time: 20–25 minutes

   A uniform plank of mass M_p and length L is placed so that it is on a support as shown below. The right side of the plank is $(3/8)L$. At the very tip of the right-hand side, a small mass m is placed.

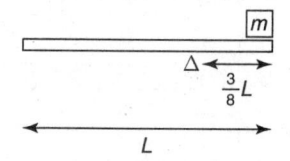

   (a) Taking the axis of rotation to be the support, draw and label arrows that represent the forces (not components) that are producing torques on the plank that might cause it to rotate. Each force in your diagram must be represented by a distinct arrow starting on, and pointing away from, the point at which the force is exerted on the plank.
   (b) Derive an expression (in terms of M_p, L, m, and physical constants as appropriate) for the net torque on the plank.
   (c) Given that the small mass m is chosen to keep the plank balanced, derive an expression for this balancing mass m.
   (d) Assume that a slightly larger mass m was chosen than in part (c) above and that the plank-mass system was then observed to rotationally accelerate from rest to an angular speed ω in a short time t. In terms of M_p, L, m, ω, and t, derive an expression of the rotational inertia of the system.
   (e) Now assume the slightly larger mass in part (d) was placed at the left-hand tip of the plank rather than the right-hand tip and the resulting angular speed in the opposite direction was measured after the same short time t as in part (d). Which of the following 3 options correctly describes how much faster the new angular speed (ω') is when compared to the angular speed (ω) measured in part (d)? Justify your reasoning.

   _____ $\omega < \omega' < 5/3\,\omega$ _____ $\omega' = 5/3\,\omega$ _____ $\omega' > 5/3\,\omega$

2. 12 points; suggested time: 25–30 minutes

   A disk of mass M and radius R begins at rest at the top of a ramp at a height H above the bottom of the ramp. The disk rolls without slipping to the bottom and has a rotational inertia of $(1/2)MR^2$.

188 AP PHYSICS 1

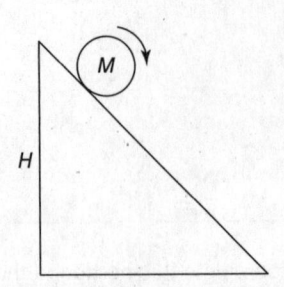

(a) Write down an expression in terms of the variables given and the instantaneous values of h, v, and ω.

(b) The energy bar charts below represent the gravitational potential energy U_g, the translational kinetic energy K_t, and the rotational kinetic energy K_r of the disk. The bar chart for when the disk is at height H has been completed for you. Draw shaded rectangles to complete the bar charts at $H/2$ and for when the disk reaches the bottom of the ramp.

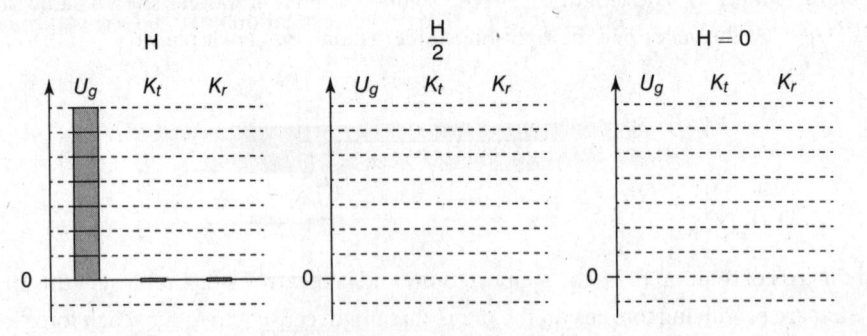

(c) Using the graph of the rotational speed of the disk provided below, derive an expression for the rotational displacement (θ_1) at t_0 and for θ_2 at time $2t_0$. Express your answer in terms of t_0, R, ω_{max}, and physical constants as appropriate.

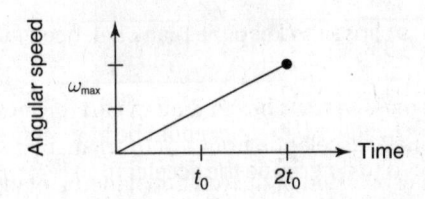

(d) On the axes provided below, sketch a graph of the angular displacement of the disk as it rolls down the ramp.

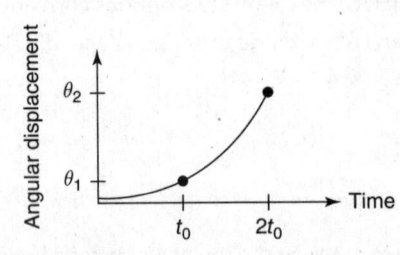

(e) A student makes the following claim: "In a real physical setting, the mechanical energy of the rolling disk is not truly conserved because of the negative work done to the system by air resistance as it rolls down the ramp." How would this claim affect your graph in part (d)? Justify your answer.

3. 10 points; suggested time: 25–30 minutes

A group of students are given the following supplies: a stopwatch, a long string, various metersticks and protractors, and a large supply of various styles of predetermined masses.

(a) Describe three short experimental procedures to determine the dependency of a simple pendulum's period of oscillation on amplitude, mass, and length. You may include a labeled diagram of your setup to help in your description. Indicate what measurements you would take and how you would take them. Include enough details so that another student could carry out your procedure.

(b) Predict the expected results of each investigation. Sketch out what the data will look like in each of the three investigations (amplitude, mass, and length).

(c) What are the common sources of error or expected deviations from ideal results that might happen during this investigation? Which of the three investigations might you expect to deviate the most from the ideal results and why?

(d) Here are some data taken from the length vs. period investigation by a student who suspects there is a correlation between the two.

Length (cm)	Period (s)	?
10	0.62	
20	0.90	
30	1.09	
40	1.28	
55	1.48	
75	1.75	
85	1.85	

Indicate which measured or calculated quantity could be plotted on a horizontal axis to yield a linear graph whose slope can be used to determine the acceleration due to gravity. Fill in these calculated values in the empty column above.

(e) On the grid below, plot the appropriate quantities to determine the acceleration due to gravity. Clearly scale and label all axes, including units, as appropriate. Draw a best-fit line to the data.

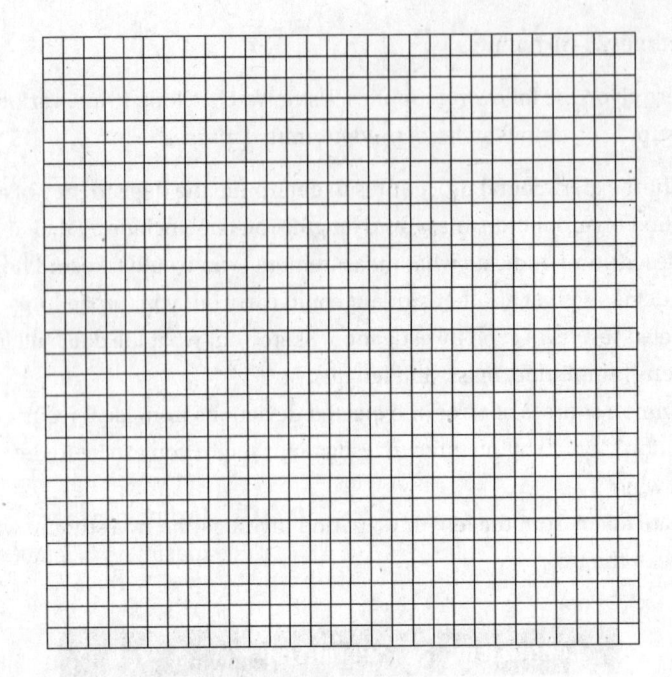

(f) Calculate an experimental value for the acceleration due to gravity using this best-fit line.

4. 8 points; suggested time: 15–20 minutes

A mass m is resting at a height h above the ground. When released, the mass can slide down a frictionless track to a loop-the-loop of radius r as shown below.

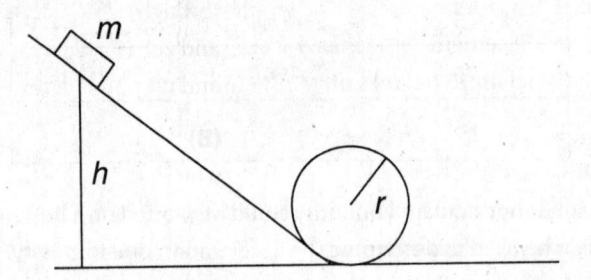

(a) If the mass m is to just barely stay on the track at the top of the loop as it slides along, estimate the relationship between the starting height h and the radius r of the loop, including any dependency on the mass m itself. Justify your estimate using qualitative reasoning beyond referencing equations.
(b) Starting with the conservation of energy, derive an expression for the velocity of the mass at the top of the loop in terms of the other variables in the problem.
(c) Using the conditions for maintaining circular motion, determine the minimum velocity required at the top of the loop.
(d) Do your equations in parts (b) and (c) agree with your qualitative reasoning in part (a)? Justify why or why not.

STOP If there is still time remaining, you may review your answers.

Answers Explained

Section I: Multiple-Choice

1. **(D)** Twice the initial vertical velocity will give twice the time in flight. Average vertical velocity will also be doubled. Displacement is the product of these two: $2 \times 2 = 4$.

   Alternatively, one could use $v^2 = v_0^2 + 2ad$, since v at the maximum height is zero. Since v_0 is doubled and then squared, the vertical displacement d must be 4 times bigger.

   Free-fall problems are independent of mass.

2. **(C)** Conservation of momentum:

   $$p_{tot} = 2(+4) + 5(-1) = +3 \text{ kg m/s} = mv = (7 \text{ kg})v$$
   $$v = +3/7 \text{ m/s}$$

   The collision is inelastic since the carts stick together.

3. **(D)** Higher altitude is strictly a function of V_Y (Projectile Y). Range is a function of both V_X and V_Y such that the angles equally above and below 45 degrees (the max angle for range) will result in equal horizontal displacement. Note that 30 degrees and 60 degrees are both 15 degrees off from 45 degrees.

4. **(B)** Since the 10 N force will start the mass moving, this must be greater than the static friction force holding the mass in place. The maximum *static* friction must be 10 N or less.

   $$F_s < 10 \text{ N}$$
   $$\mu N < 10 \text{ N}$$
   $$\mu mg < 10 \text{ N}$$
   $$\mu(5)(10) < 10$$
   $$\mu < 1/5$$

5. **(B)** The object is moving, so the velocity is not zero. The object is not accelerating, so velocity is constant.

6. **(C)** In general:

   $$y(t) = y_0 - v_{y0} - \frac{1}{2}gt^2$$
   $$x(t) = v_{x0} t$$

   Solve for t in the last equation. Then plug back into the first equation and substitute in $y_0 = h$, $v_{y0} = 0$, and $v_{x0} = v_0$:

   $$y(x) = h - (0)(x/v_0) - \frac{1}{2}g(x/v_0)^2$$
   $$= h - gx^2/(2v_0^2)$$

7. **(C)** Maximum acceleration allows you to determine maximum displacement:

   $$ma = kA$$

   Knowing the amplitude allows you to determine easily the maximum speed via energy conservation:

   $$\frac{1}{2}kA^2 = \frac{1}{2}mv^2$$

8. **(B)** Maximum velocities happen when going through the equilibrium point (zero acceleration and zero displacement): all kinetic energy (*KE*) and no potential energy (*PE*).

9. **(B)** Momentum is conserved during the collision, which enables us to solve for an initial upward velocity of the combination. Then energy conservation can be used to relate the height to that initial upward velocity. Choice (A) cannot be used because some unknown amount of mechanical energy will be lost by the bullet embedding itself in the wood.

10. **(A)** Constant velocity means no acceleration, so $\vec{F}_{net} = 0$.

    $$F_{push} - F_{friction} = 0$$
    $$F_{push} - \mu_k N = F_{push} - \mu_k mg = 0$$
    $$F_{push} = \mu_k mg = (0.1)(10 \text{ kg})(10 \text{ m/s}^2) = 10 \text{ N}$$

11. **(D)** First find the acceleration of the system:

$$\vec{F}_{net} = m\vec{a}$$
$$10 \text{ N} = (4 + 1 \text{ kg})a$$
$$a = 2 \text{ m/s}^2$$

The contact force, P, is the only force felt by the 1-kg mass:

$$\vec{F}_{net} = m\vec{a}$$
$$P = (1 \text{ kg})(2 \text{ m/s}^2) = 2 \text{ N}$$

12. **(D)** $P = \text{work}/t = Fd/t = mgh/t$

13. **(A)** Impulse can be found by Ft if the details of the force are known or, alternatively, by:

$$\text{Impulse} = \Delta p = (1,500 \text{ kg})(0 \text{ m/s} - 25 \text{ m/s})$$
$$= -37,500 \text{ N} \cdot \text{s}$$

Ignore the minus sign as the question asks about magnitude.

14. **(A)** When the block is at the maximum height, static friction is obtained:

$$\vec{F}_{net} = mg \sin \theta - \mu N = 0$$
$$mg \sin \theta - \mu(mg \cos \theta) = 0$$
$$\mu = (\sin \theta)/(\cos \theta) = \tan \theta$$

15. **(C)** Doubling the period requires a quadrupling of length:

$$T = 2\pi(l/g)^{1/2}$$

16. **(B)** Impulse = area of F vs. T graph = Δp
$$= mv_f - 0$$

Solving for $v_f = (\text{area under graph})/m$

17. **(C)** Conservation of energy:

$$\frac{1}{2}mv^2 = mg\Delta h$$
$$\frac{1}{2}v^2 = g\Delta h = (10 \text{ m/s}^2)(2.25 \text{ m} - 0.75 \text{ m})$$
$$v = (30)^{1/2} = 5.5 \text{ m/s}$$

18. **(D)** Gravitational field strength halfway between any two equal masses is always zero as each contributes oppositely directed gravitational fields.

19. **(D)** Conservation of energy while on the frictionless hill can give the speed at the top of the 10-m hill:

$$\frac{1}{2}mv^2 = KE = mg\Delta h$$

Then impulse equals change in momentum can be used since the final momentum must be zero.

20. **(A)**

$$\text{Mechanical energy} = mgh + \frac{1}{2}mv^2$$

$$\text{Energy "lost" or "gained"} = mgH_2 + \frac{1}{2}mv_2^2$$
$$- \left(mgH_1 + \frac{1}{2}mv_1^2\right)$$

Work by forces other than gravity = change in energy

21. **(B)** Energy conservation:

$$\frac{1}{2}kx^2 = mgh$$

Solving for h:

$$h = kx^2/2mg$$

Doubling x will quadruple the height, whereas all other factors will only double the height.

22. **(B)** Circular motion at the top:

$$F_{net} = mg + N = mv^2/r$$

The lowest speed will be when there is no normal force ($N = 0$):

$$mg = mv^2/r$$
$$v_{top} = (gr)^{1/2}$$

Kinetic energy at the bottom must give both this speed and potential energy to gain $2r$ in height:

$$E_{bottom} = E_{top}$$
$$\frac{1}{2}mv^2 = gm(2r) + \frac{1}{2}mv_{top}^2$$
$$\frac{1}{2}mv^2 = gm(2r) + \frac{1}{2}m(gr)$$

Solving for v:

$$v = (5gr)^{1/2}$$

23. **(C)** Since the force shown has components both in the same direction as the velocity (causing it to speed up) and at right angles to the velocity (causing it to turn toward the bottom of the page), the particle will both speed up and turn.

24. **(C)** Since the elevator is moving downward at constant speed, the y-component of velocity should be a constant negative value. On the other hand, the horizontal acceleration of the boat will cause an increasingly positive x-component of velocity.

25. **(B)** Impulse is the area under the curve. For the first three seconds, this amounts to

$$½ (6)(3) = 9 \text{ Ns}$$

This impulse will increase the momentum by 9 kg m/s. Since the object has a mass of 9 kg, this indicates an increase in speed of 1 m/s.

26. **(A)** Work done is equal to the change in kinetic energy:

$$K_f - K_i = 0 - ½ (3m)v^2 = -1.5 \, mv^2$$

27. **(C)** Circular orbits have universal gravity acting as the centripetal force, which causes the centripetal acceleration of v^2/r:

$$F_c = ma_c$$
$$GM_eM/r^2 = mv^2/r$$

Since all other values are constant, we see here v^2 is proportional to $1/r$.
So an orbit that has twice the radius will have half the v^2.
Kinetic energy varies a v^2:

$$K = ½ \, mv^2$$

Thus, the higher orbital will have half the kinetic energy of the lower orbit.

28. **(C)** A change in rotational speed can either be accomplished via an interaction with an external torque or by a change in rotational inertia, which keeps angular momentum conserved in an isolated system by adjusting the angular speed:

$$I\omega = \text{constant}$$

Since the earthquake is not an external force to the Earth, the shifting of mass to different radial values must have changed the rotational inertia of the planet. This is similar to a diver adjusting their spin rate by extending or tucking their body in as they drop into the water.

29. **(C)** Weight is greater for the wider column, while pressure is the same. Although the wider column has a greater volume of water and therefore more mass, the fact that the height remains constant means any unit area on the bottom of the column will experience the same force/area (pressure).

30. **(A)** The fluid's density is raised. The buoyancy force that lifts the boat upward is proportional to the weight of the fluid displaced. The boat will rise or sink until the buoyant force and the boat's weight cancel. Raising the fluid's temperature will usually make the fluid less dense, requiring a greater displacement of water (i.e., the boat will ride lower). Raising the boat's density will increase its weight, requiring a greater displacement of water. Finally, shifting the cargo around will have no effect as the weight and shape of the boat remain unaffected.

31. **(B)** It will rise on the right-hand side and lower on the left-hand side. Blowing over the right-hand side will lower the pressure at that height:

$$1/2 \, (\rho)v + (\rho)gh + P = \text{constant}$$

Remember that h remains the same while v is raised, forcing P lower. This causes an imbalance in pressure between the right-hand side (lower pressure) and left-hand side (higher pressure). This difference in pressure causes a net force pushing the fluid upward on the right and downward on the left.

32. **(B)** Find time to cross the river using only perpendicular components:

$$D = V_y t$$
$$240 = 8t$$
$$t = 30 \text{ s}$$

Next find the downstream distance using the parallel component:

$$D = V_x t$$
$$D = (6 \text{ m/s})(30 \text{ s}) = 180 \text{ m}$$

194   AP PHYSICS 1

33. **(C)** Work is the area under the curve:

$$(4\text{ N})(6\text{ m}) + (2\text{ N})(4\text{ m}) = 24\text{ J} + 8\text{ J} = 32\text{ J}$$

34. **(B)** The bathroom scale reads the normal force:

$$F_{net} = N - mg = ma$$
$$0.5\text{ m/s}^2 = 0.05g$$
$$N - mg = m(0.05g)$$
$$N = m(1.05g)$$

The reading will be 5% too high.

35. **(D)** Balanced means the torques are equal and opposite:

$$50\text{ kg} \cdot g \cdot 1.2 = 70\text{ kg} \cdot g \cdot x$$
$$x = 6/7 = 0.86\text{ m}$$

36. **(C)** Both objects start with the same potential energy. However, the rolling object must use some of that potential energy for rotational energy, leaving less for linear kinetic energy. Therefore, the rolling object moves more slowly down the hill.

37. **(C)** Universal gravity (and the gravitational field) are $1/r^2$ laws; doubling r will quarter the field. $(1/4)g = (1/4)(9.8\text{ N/kg})$. Note that N/kg = m/s^2.

38. **(A)** First, convert RPMs to rad/s:

150 rev/min × (2π/1 rev) × (1 min/60 s) = 5π rad/s

acceleration = change in velocity/time
= (0 − 5π rad/s)/(2 s) = −2.5π rad/s^2

39. **(D)** $\frac{1}{2}kx^2$ of work is delivered to the mass while uncompressing, followed by $-\frac{1}{2}kx^2$ done while the mass extends the spring outward, totaling to 0 net work for the entire expansion. Alternatively, think about the velocity being zero at the beginning and end of that single oscillation. No change in kinetic energy occurs; therefore, no net work is done.

40. **(D)** No external torque means momentum must be conserved.

Section II: Free-Response

1. (a) 1 point for two downward gravitational arrows.
   1 point for $M_p g$ located at the center of mass of the rod ($(1/8)L$ to the left of the support) and for Mg being at the far-right-hand end.

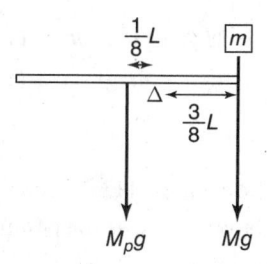

   (b) 1 point for opposing signs on the two torques.
   1 point for the proper expression for net torque.

   $$\tau_{net} = M_p g(1/8)L \sin 90° - Mg(3/8)L \sin 90°$$
   $$= (gL/8)(M_p - 3M)$$

   (c) 1 point. Set $\tau_{net} = 0$ and solve for M:

   $$M = M_p/3$$

   (d) 1 point for an expression for angular acceleration:

   $$\alpha = (\omega - 0)/t = \omega/t$$

   1 point for using $\tau_{net} = I\alpha$ to find momentum of inertia:

   $$(gL/8)(M_p - 3M) = I\alpha = I\omega/t$$
   $$I = (gLt/8\omega)(M_p - 3M)$$

   (e) 1 point for choosing $\omega' > 5/3 \, \omega$.
   1 point for including a justification. If the mass of the plank itself is ignored, the new torque would be 5/3 times as great due to the increased lever arm. However, since the center of mass and hence the torque of the mass of the plank itself opposed the initial torque but works with the new torque, the net torque will be much greater:

   $$(gL/8)(M_p + 5M) > (gL/8)(M_p - 3M)$$

   Note that the original expression (right-hand side) is only slightly negative under the conditions described in part (d).

   1 point for discussing the rotational inertia. Note that the rotational inertia in this configuration will be slightly greater. This increase is only $M(L/4)^2$. Although this would decrease the angular acceleration on its own, it is not a large enough to offset the increase due to the much larger torque.

2. (a) 1 point for proper expression of total mechanical energy:

$$\text{mechanical energy} = Mgh + (1/2)Mv^2 + (1/2)I\omega^2$$

Putting in the criteria for rolling without slipping ($\omega = v/R$) and the given rotational inertia for a sphere produces:

$$\text{mechanical energy} = Mgh + (1/2)Mv^2 + (1/2)(2/5MR^2)(v/R)^2$$
$$= Mgh + (1/2)Mv^2 + (1/5)Mv^2$$

1 point for a final expression in this form.

(b) From the expression above, we see the rotational kinetic energy (last term above) is in a 2:5 ratio with the linear kinetic energy (second term above). Thus, as the potential energy is lost to the kinetic energies, it must be split between them in this manner.

1 point for U_g being 3.5 units in the second chart and 0 units in the final chart.
2 points for K_t being 2.5 units and K_r being 1 unit in the second chart.
2 points for K_t being 5 units and K_r being 2 units in the second chart.

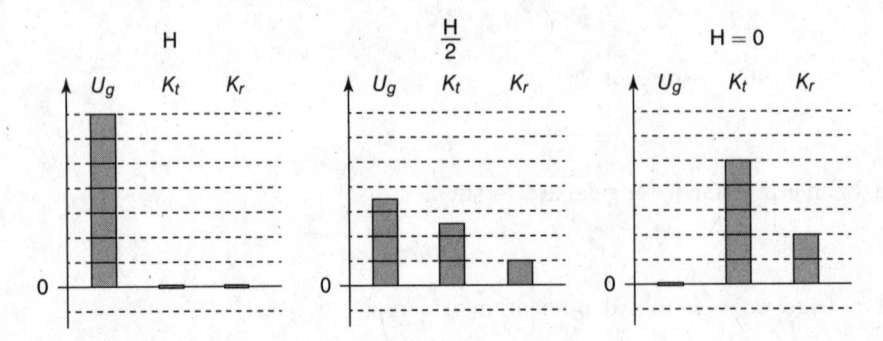

(c) Angular displacement is the area under the curve.

1 point for:

$$\theta_1 = (1/2)(\omega_{max}/2)(t_0) = (1/4)\omega_{max}t_0$$

1 point for:

$$\theta_2 = (1/2)(\omega_{max})(2t_0) = \omega_{max}t$$

(d) 1 point for an upward curving graph.
1 point for being 4 times higher at $2t_0$.

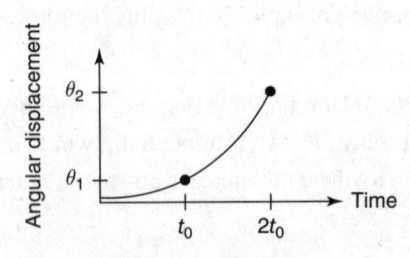

(e) 1 point for explaining. Although the graph would still be an upward curve, it would not be as steep because the acceleration would be slightly reduced by the friction. The actual angular displacement would start off in the same manner but would become increasingly reduced as the sphere picked up speed.

3. (a) 1 point. This is really three separate investigations. In each case, students should measure the period repeatedly and then take the mean value.

   1 point for procedures similar to these:
   - Amplitude versus period. Take a period measurement for 5 to 10 different amplitudes while keeping the mass and length the same. The amplitude can be controlled by pulling out the string-mass combo to a certain angle as measured by the protractor.
   - Mass versus period. Take a period measurement for 5 to 10 different masses while keeping the amplitude and length constant.
   - Length versus period. Take a period measurement for 5 to 10 different lengths while keeping the mass and amplitude constant. Length is measured from the pivot point to the center of mass.

   (b) 1 point for stating that no correlation is expected for amplitude and mass variations.

   1 point for a nonlinear relationship is expected for length and period:

   $$T_p = 2\pi(L/g)^{1/2}$$

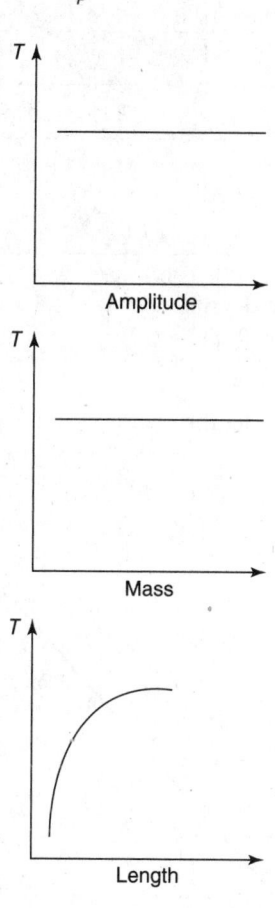

   [Graph: T vs Amplitude — horizontal line]

   [Graph: T vs Mass — horizontal line]

   [Graph: T vs Length — square-root shaped curve]

   (c) 1 point. Beyond the usual random errors of measurement (especially when using a timer but minimized by taking the median value of a few trials each time), one predictable deviation is in the amplitude investigation. Pendulums actually behave as simple harmonic oscillators only under the conditions of small angles (small enough that $\sin\theta$ is approximately θ). A large enough amplitude will require the pendulum to oscillate at larger angles. This means that at large amplitudes, one can predict the results to deviate from the expected as the gravitational force no longer acts as a simple restorative force.

One other possible source of systematic error would be in the mass investigation. As various masses are swapped out on a fixed length of string, the students may inadvertently be changing the length of the string when adding different-sized masses. The length of the pendulum is from the pivot point to the center of mass. If the students do not compensate for this by shortening the string when adding larger masses, they may see an artificial relationship at higher masses in their graph of mass versus period.

(d) 2 points. To obtain a linear graph, you must plot T^2 versus L:

$$T^2 = (4\pi^2/g)L$$

The slope of this plot would then be equal to $4\pi^2/g$. Alternatively, you could plot T versus $\sqrt{L}$.

Length (cm)	Period (s)	T^2 (s²)
10	0.62	0.38
20	0.90	0.81
30	1.09	1.19
40	1.28	1.64
55	1.48	2.19
75	1.75	3.06
85	1.85	3.42

(e) 1 point for graph, including labels and units.
1 point for the line of best fit.

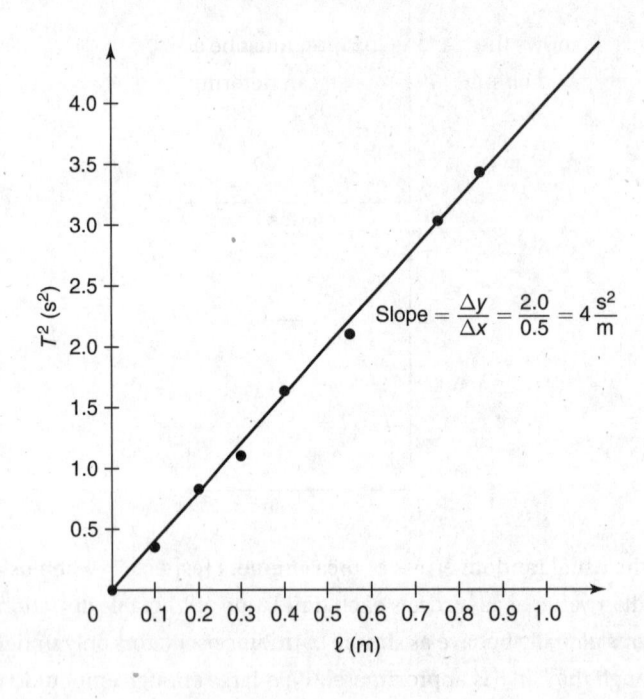

(f) 1 point. The measured slope is 4 s²/m. Setting this equal to $4\pi^2/g$, from part (d), and solving for g gives:

$$g = 4\pi^2/4 = 9.9 \text{ m/s}^2$$

4. (a) 2 points. The starting height must be above the value of $2r$. Starting from rest, a starting height of $2r$ would result in a velocity of zero at the top of the inner loop, which is an insufficient speed to maintain circular motion. The additional height above $2r$ must have enough additional potential energy in order to give the mass enough kinetic energy to maintain circular motion at the top.

(b) 1 point for correctly stating the energy at the top of the loop and at the very beginning:

$$E_{top} = mg(2r) + (1/2)mv^2 = E_{beginning} = mgh$$

1 point for solving for v:

$$v^2 = 2g(h - 2r)$$
$$v = (2g(h - 2r))^{1/2}$$

(c) 1 point for identifying that the minimum centripetal force is provided by the force of gravity alone:

$$F_c = mg$$

1 point for using the centripetal acceleration formula and finding v:

$$mg = ma_c = mv^2/r$$
$$v = (gr)^{1/2}$$

(d) 1 point for setting these two expressions for velocity equal to each other and solving for h:

$$(2g(h - 2r))^{1/2} = (gr)^{1/2}$$
$$2g(h - 2r) = gr$$
$$2(h - 2r) = r$$
$$h = 2.5r$$

1 point. Yes, this result shows the starting location must be an additional $0.5r$ above the $2r$ value. By using conservation of energy and circular motion, we can determine the amount of kinetic energy needed at the top of the loop is $0.5mgr$.

Test Analysis

Practice Test 1

Section I: Multiple-Choice

Note that the questions requiring two answers are to be graded as completely correct (1 point) or incorrect (0 points).

Number correct (out of 40) = $\dfrac{}{\text{Multiple-Choice Score}}$

Section II: Free-Response

Partial credit is awarded for any correct responses within an individual free-response question.

Question 1 = $\dfrac{}{\text{(out of 10)}}$

Question 2 = $\dfrac{}{\text{(out of 12)}}$

Question 3 = $\dfrac{}{\text{(out of 10)}}$

Question 4 = $\dfrac{}{\text{(out of 8)}}$

Total = $\dfrac{}{\text{(out of 40)}}$

Final Score

$\dfrac{}{\text{Multiple-Choice Score}} + \dfrac{}{\text{Free-Response Score}} = \dfrac{}{\text{Total (out of 80)}}$

Final Score Range*

Final Score Range	AP Score
60–80	5
44–59	4
32–43	3
20–31	2
0–19	1

*Note: The guidelines above are based on the released scores for past AP Physics 1 exams. Actual score ranges vary from year to year and are determined by the College Board each year. Thus, the ranges shown are approximate.

ANSWER SHEET
Practice Test 2

1. Ⓐ Ⓑ Ⓒ Ⓓ
2. Ⓐ Ⓑ Ⓒ Ⓓ
3. Ⓐ Ⓑ Ⓒ Ⓓ
4. Ⓐ Ⓑ Ⓒ Ⓓ
5. Ⓐ Ⓑ Ⓒ Ⓓ
6. Ⓐ Ⓑ Ⓒ Ⓓ
7. Ⓐ Ⓑ Ⓒ Ⓓ
8. Ⓐ Ⓑ Ⓒ Ⓓ
9. Ⓐ Ⓑ Ⓒ Ⓓ
10. Ⓐ Ⓑ Ⓒ Ⓓ
11. Ⓐ Ⓑ Ⓒ Ⓓ
12. Ⓐ Ⓑ Ⓒ Ⓓ
13. Ⓐ Ⓑ Ⓒ Ⓓ
14. Ⓐ Ⓑ Ⓒ Ⓓ
15. Ⓐ Ⓑ Ⓒ Ⓓ
16. Ⓐ Ⓑ Ⓒ Ⓓ
17. Ⓐ Ⓑ Ⓒ Ⓓ
18. Ⓐ Ⓑ Ⓒ Ⓓ
19. Ⓐ Ⓑ Ⓒ Ⓓ
20. Ⓐ Ⓑ Ⓒ Ⓓ
21. Ⓐ Ⓑ Ⓒ Ⓓ
22. Ⓐ Ⓑ Ⓒ Ⓓ
23. Ⓐ Ⓑ Ⓒ Ⓓ
24. Ⓐ Ⓑ Ⓒ Ⓓ
25. Ⓐ Ⓑ Ⓒ Ⓓ
26. Ⓐ Ⓑ Ⓒ Ⓓ
27. Ⓐ Ⓑ Ⓒ Ⓓ
28. Ⓐ Ⓑ Ⓒ Ⓓ
29. Ⓐ Ⓑ Ⓒ Ⓓ
30. Ⓐ Ⓑ Ⓒ Ⓓ
31. Ⓐ Ⓑ Ⓒ Ⓓ
32. Ⓐ Ⓑ Ⓒ Ⓓ
33. Ⓐ Ⓑ Ⓒ Ⓓ
34. Ⓐ Ⓑ Ⓒ Ⓓ
35. Ⓐ Ⓑ Ⓒ Ⓓ
36. Ⓐ Ⓑ Ⓒ Ⓓ
37. Ⓐ Ⓑ Ⓒ Ⓓ
38. Ⓐ Ⓑ Ⓒ Ⓓ
39. Ⓐ Ⓑ Ⓒ Ⓓ
40. Ⓐ Ⓑ Ⓒ Ⓓ

Practice Test 2

Section I: Multiple-Choice

TIME: 80 MINUTES
40 QUESTIONS

DIRECTIONS: Each of the questions or incomplete statements below is followed by four suggested answers or completions. Select the one that is best in each case. You have 80 minutes to complete this portion of the test. You may use a calculator and the information sheets provided in the appendix.

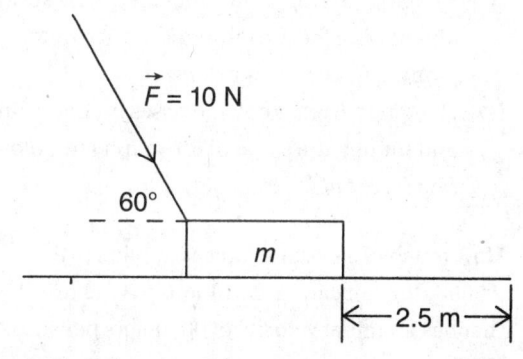

1. A block ($m = 1.5$ kg) is pushed along a frictionless surface for a distance of 2.5 meters, as shown above. How much work has been done if a force of 10 newtons makes an angle of 60 degrees with the horizontal?

   (A) Zero
   (B) 12.5 J
   (C) 21.6 J
   (D) 25 J

2. What is the instantaneous power due to the gravitational force acting on a 3-kilogram projectile the instant the projectile is traveling with a velocity of 10 meters per second at an angle of 30 degrees above the horizontal?

   (A) 300 W
   (B) 150 W
   (C) −150 W
   (D) −300 W

3. A hockey puck of an unknown mass is sliding along ice that can be considered frictionless with a velocity of 10 meters per second. The puck then crosses over onto a rough floor that has a coefficient of kinetic friction equal to 0.2. How far will the puck travel before friction stops it?

   (A) 2.5 m
   (B) 5 m
   (C) 25 m
   (D) Depends on the mass

4. A dart is placed onto a spring. The spring is stretched a distance x. By what factor must the spring's elongation be changed so that the maximum kinetic energy given to the dart is doubled?

   (A) 1/2
   (B) 2
   (C) 4
   (D) $\sqrt{2}$

5. A 10-kilogram projectile is launched at a 60° angle to the ground, with a velocity of 200 m/s. Neglect air resistance. The ground is level.

   Compare this projectile with a 5-kilogram projectile launched under the same conditions but at a 30° angle. The 5-kilogram projectile will

   (A) go higher up and farther along the ground
   (B) go equally high and equally far along the ground
   (C) neither go as high nor as far along the ground
   (D) not go as high but go equally far along the ground

6. A uniform rod of mass M and length L is rotated about its center as shown below. Given that the rotational inertia of a rod about its center is $(1/12)ML^2$, determine the net torque experience by the rod as it rotates from an angular speed of ω_1 to ω_2 in 0.5 seconds.

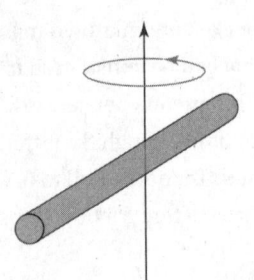

   (A) $(\omega_2 - \omega_1)ML^2/12$
   (B) $(\omega_2 - \omega_1)ML^2/6$
   (C) $\omega_2\omega_1 ML^2$
   (D) $(\omega_2/\omega_1)ML^2/12$

7. An object of mass m rests on top of a spring that has been compressed by x meters. The force constant for this spring is k. The mass is not attached to the spring and will shoot upward when the spring is uncompressed. When released, how high will the mass rise?

   (A) $mg - kx$
   (B) kx^2/mg
   (C) $(kx^2/2mg) - x$
   (D) $(k/m)^{1/2}x$

8. Which of the following is the best method for finding a spring's force constant k?

   (A) Hanging a known mass on the spring and dividing the weight by the length of the spring
   (B) Hanging several known masses on the spring, taking the average value of the mass, and dividing by the average length of the spring
   (C) Hanging several known masses on the spring and finding the area under the curve after plotting force versus extension
   (D) Hanging several known masses on the spring and finding the slope of the graph after plotting force versus extension

9. How much of a braking force is applied to a 2,500-kilogram car on the Moon ($g = 1.6$ m/s^2) that has an initial velocity of 30 meters per second if the car is brought to a stop in 15 seconds?

   (A) 5,000 N
   (B) 6,000 N
   (C) 8,000 N
   (D) 25,000 N

10. A 1-kilogram object is moving to the right with a velocity of 6 meters per second. It collides with and sticks to a 2-kilogram mass, which is also moving to the right, with a velocity of 3 meters per second. What happens to the total kinetic energy during this collision?

   (A) The kinetic energy is conserved because the collision is elastic.
   (B) The kinetic energy is conserved even though the collision is not elastic.
   (C) Some kinetic energy is lost during the collision even though total momentum is conserved.
   (D) Some kinetic energy is lost during the collision because of the elastic nature of the collision.

11. A ball with a mass of 0.2 kilogram strikes a wall with a velocity of 3 meters per second. It bounces straight back with a velocity of 1 meter per second. What was the magnitude of the impulse delivered to this ball?

   (A) 0.2 kg · m/s
   (B) 0.4 kg · m/s
   (C) 0.6 kg · m/s
   (D) 0.8 kg · m/s

12. Which of the following is an equivalent expression for the maximum velocity attained by a mass m oscillating horizontally along a frictionless surface? The mass is attached to a spring with a force constant k and has an amplitude of A.

   (A) Ak/m
   (B) $A(k/m)^{1/2}$
   (C) mg/kA
   (D) A^2k/m

13. A 0.5-kilogram mass is attached to a spring with a force constant of 50 newtons per meter. What is the total energy stored in the mass-spring system if the mass travels a distance of 8 cm in one cycle?

   (A) 0.5 J
   (B) 0.01 J
   (C) 0.04 J
   (D) 0.08 J

14. Which of the following expressions is equivalent to the magnitude of the escape velocity in terms of the magnitude of the orbital velocity v for a spacecraft?

   (A) $2v$
   (B) v
   (C) $4v$
   (D) $\sqrt{2}\,v$

15. Which of the following graphs correctly shows the relationship between magnitude of gravitational force and distance between two masses?

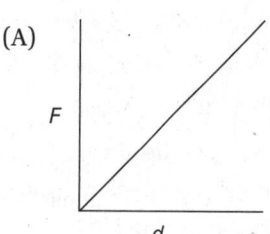

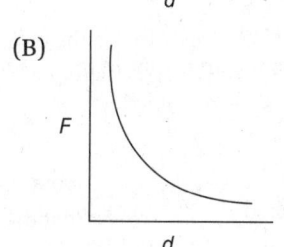

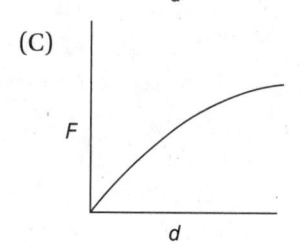

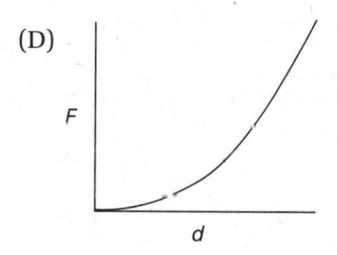

   (A)
   (B)
   (C)
   (D)

16. Which of the following statements is true regarding why the net work done in lifting a book from the ground to the top of a shelf is zero?

    (A) Air resistance (friction) took away any increase in kinetic energy.
    (B) Gravitational potential energies are always negative.
    (C) The force used to lift the book is a conservative force.
    (D) The change in kinetic energy of the book is zero.

17. Which of the following best describes the forces present as a brick is sliding along the horizontal ground and coming to a halt?

    (A) The net force is equal to the force of friction.
    (B) The inertia of the brick is supplied by the original push that got the brick moving.
    (C) The forces on the brick are in equilibrium.
    (D) The brick is pushing on the air in front of it harder than the air is pushing back on the brick.

18. A projectile is launched at a 30° angle to the ground with a velocity of 200 m/s. What is its speed at its maximum height?

    (A) 9.8 m/s
    (B) 100 m/s
    (C) 173 m/s
    (D) 200 m/s

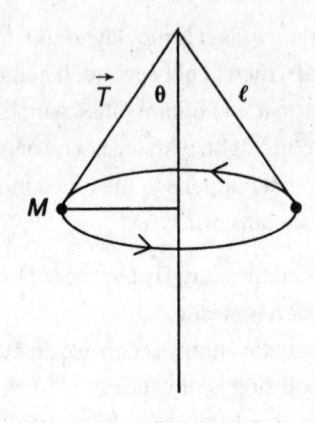

19. A conical pendulum consists of a mass m attached to a light string of length L. The mass swings around in a horizontal circle, making an angle θ with the vertical as shown above. What is the magnitude of tension, T, in the string?

    (A) $mg/\cos\theta$
    (B) $mg\cos\theta$
    (C) $mg/\sin\theta$
    (D) $mg\sin\theta$

20. The magnitude of the one-dimensional momentum of a 2-kilogram particle obeys the relationship $p = 2t + 3$. What was the velocity of the particle at $t = 1$ second?

    (A) 5 m/s
    (B) 2 m/s
    (C) 1 m/s
    (D) 2.5 m/s

21. An object is experiencing a nonzero net force. Which of the following statements is most accurate?

    (A) The linear and angular momentums of the object are both definitely changing.
    (B) Although the linear momentum of the object is definitely changing, the angular momentum may not be.
    (C) Although the angular momentum of the object is definitely changing, the linear momentum may not be.
    (D) Neither the linear momentum nor the angular momentum is definitely changing.

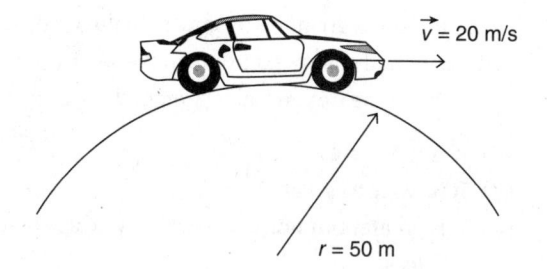

22. A car with a 500-newton driver goes over a hill that has a radius of 50 meters as shown above. The velocity of the car is 20 meters per second. What are the approximate force and the direction that the car exerts on the driver?

(A) 900 N, up
(B) 400 N, down
(C) 100 N, up
(D) 500 N, up

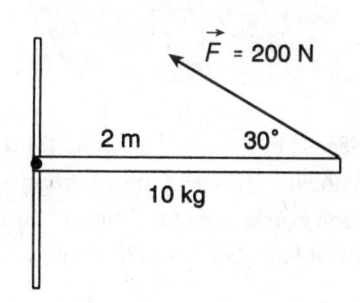

23. What is the net torque acting on the pivot supporting a uniform stationary 10-kilogram beam 2 meters long as shown above?

(A) 198 N · m
(B) −198 N · m
(C) −102 N · m
(D) 102 N · m

Use the graph below to answer the following four questions (24–27):

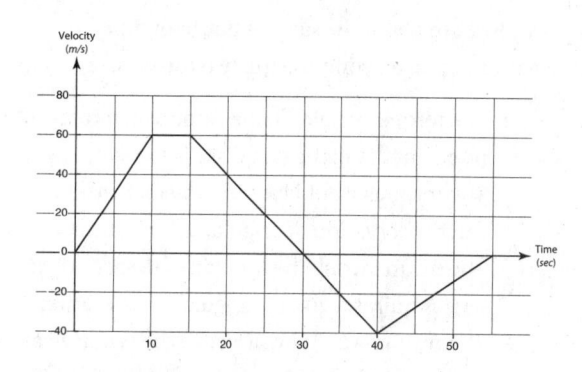

24. Which of the following statements best describes the acceleration over the 55 seconds of time graphed above?

(A) Increasing, constant, decreasing, increasing
(B) Positive, zero, negative, positive
(C) Positive until the 30 second mark, then negative for the rest of the trip
(D) The net acceleration is approximately 500 m/s/s for the entire trip.

25. What is the average acceleration for the 55 seconds of the graph?

(A) Approximately 500/55 m/s/s
(B) Approximately 5/4 m/s/s
(C) Approximately −100/25 m/s/s
(D) Approximately 0 m/s/s

26. How far does the car travel over the 55 seconds of the graph?

(A) 1,550 m
(B) 550 m
(C) 0 m
(D) Cannot be determined

27. What is the instantaneous acceleration at the 40 second mark?

(A) 0 m/s/s
(B) −4/3 m/s/s
(C) −10/3 m/s/s
(D) −4 m/s/s

28. An old record player could bring a disk up to its 45 RPM speed in less than a second. If the same size disk can also be brought up to a speed of 75 RPM in about the same amount of time on another player, compare the two torques.

   (A) The torques would be the same as the rotational inertia of the two disks is the same.
   (B) The torques would be the same because of the conservation of angular momentum.
   (C) The torque would be larger in the second case as it requires a greater angular acceleration.
   (D) The torque would be larger in the second case as it entails both a larger force and a larger lever arm.

Questions 29–31 refer to the following information:

Two small, identical metal spheres are projected at the same time from the same height by two identical spring guns. Each gun provides the same push on its sphere. However, one sphere is projected vertically upward while the other sphere is projected horizontally. The speed of each projectile as it emerges from the gun is the same. Frictional losses are negligible.

29. How does the speed of the vertically launched sphere compare to the speed of the horizontally launched sphere as they each hit the floor?

   (A) It is the same.
   (B) It is twice as great.
   (C) It is greater but not necessarily twice as great.
   (D) It is less.

30. How does the time required for the vertically projected sphere to hit the floor compare with that for the horizontally projected sphere?

   (A) It is the same.
   (B) It is twice as great.
   (C) It is greater but not necessarily twice as great.
   (D) It is less.

31. How does the work done by gravity to the vertically launched sphere compare to the work done by gravity to the horizontally launched sphere?

   (A) It is the same.
   (B) It is twice as great.
   (C) It is greater but not necessarily twice as great.
   (D) It is less.

Questions 32 and 33 refer to this picture of a uniform cube of volume V floating at equilibrium in a liquid as shown:

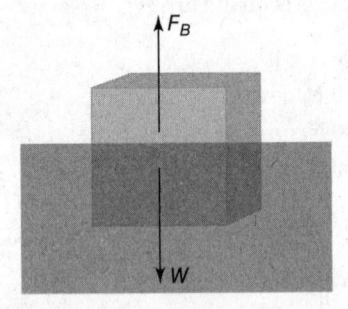

32. If the density of the liquid is ρ_L and the density of the cube if ρ_c, which of the following is the correction expression of the mass of the cube? The cube is half submerged in the liquid.

   (A) $\rho_L V$
   (B) $\rho_c V/2$
   (C) $\rho_L V/2$
   (D) $(\rho_L - \rho_c)V$

33. If the cube is submerged just under the surface of the liquid and then released, what will be the initial upward acceleration experienced by the cube?

   (A) $(\rho_L - \rho_c)Vg$
   (B) $(\rho_L - \rho_c)Vg/\rho_c$
   (C) $((\rho_L - \rho_c)/\rho_c)^{1/2}g$
   (D) $(\rho_L - \rho_c)g/\rho_c$

Questions 34 and 35 refer to the following information:

A spinning disk (with rotational inertia I) is recorded to spin for 3 seconds.

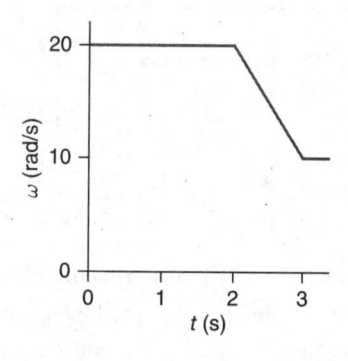

34. How many radians of rotation did the disk experience during these 3 seconds?

   (A) 60
   (B) 55
   (C) 45
   (D) 15

35. What is the change in angular momentum experienced by the disk during each of the 3 seconds of rotation?

	0–1 seconds	1–2 seconds	2–3 seconds
(A)	0	0	$-10I$
(B)	$20I$	$20I$	$-15I$
(C)	0	0	$-10I/3$
(D)	$-10I/3$	$-10I/3$	$-10I/3$

36. A mass M at the end of a string is spun in a circle of radius R and a constant speed V. Compare the tension in the line when the circle is in a horizontal plane to that of the tensions at the top and bottom of a vertical plane.

	Horizontal	Vertical-top	Vertical-bottom
(A)	MV^2/R	MV^2/R	MV^2/R
(B)	MV^2/R	$MV^2/R - Mg$	$MV^2/R - Mg$
(C)	MV^2/R	$MV^2/R + Mg$	$MV^2/R - Mg$
(D)	MV^2/R	$MV^2/R - Mg$	$MV^2/R + Mg$

37. An astronaut wakes up inside an accelerating rocket ship in deep space. Using her rubber band of unknown spring constant with attached (unknown) mass, she holds it up against the acceleration and notices the mass stretches the band downward by X meters. She then pulls it down an additional 5 cm and times the oscillation up and down to be T seconds. The acceleration of her rocket ship in terms of her measurements is

   (A) X/T^2
   (B) KX/m
   (C) $X(2\pi/T)^2$
   (D) $0.2/T^2$

38. A woman standing on a scale in an elevator notices that the scale reads her true weight. From this, she may conclude that

   (A) the elevator must be at rest
   (B) the elevator must be accelerating precisely at 9.8 m/s^2
   (C) the elevator must be moving upward
   (D) the elevator must not be accelerating

39. A planet has half the mass of Earth and half the radius. Compared with the acceleration due to gravity at the surface of Earth, the acceleration due to gravity at the surface of this planet is

   (A) the same
   (B) halved
   (C) doubled
   (D) quadrupled

40. Which of the following materials will exert the highest pressure on its bottom surface?

   (A) 1 liter of water in a puddle on the ground
   (B) 1 liter of water, frozen into a cube on the ground
   (C) 1 liter of water in a tall, thin, vertical tube
   (D) 1 liter of water, frozen into a cube, floating in liquid water

Section II: Free-Response

TIME: 100 MINUTES
4 QUESTIONS

DIRECTIONS: You have 100 minutes to complete this portion of the test. You may use a calculator and the information sheets provided in the appendix.

1. 10 points; suggested time: 20–25 minutes

   A rolling uniform sphere of mass M and radius R (rotational inertia $= (2/5)MR^2$) rolls up a curved incline as pictured below with an initial speed of v_0. At each instant, the incline can be described as having an instantaneous angle with the horizontal, θ, as indicated. This value is getting bigger as the sphere rolls upward. At all times, the sphere is rolling without slipping. The frictional force exerted on the disk is F_f.

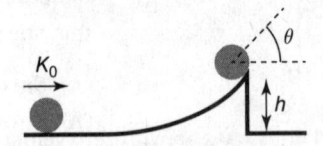

   (a) At an instant along its upward roll, draw and label arrows that represent the forces (not components) that are producing torques on the plank that might cause it to rotate. Each force in your diagram must be represented by a distinct arrow starting on and pointing away from the point at which the force is exerted on the plank. Specify if the torque induced is a clockwise or counterclockwise one.

   (b) Derive an expression (in terms of M, R, θ, F_f, and physical constants as appropriate) for the net torque on the sphere about its center of mass.

   (c) Starting with a statement of conservation of energy, derive an expression for the height of the rolling sphere in terms of M, R, h, θ, v_0, and physical constants as appropriate.

   (d) At some point, the ball will reach its highest point. Based on your expression derived in part (c), determine an expression for this highest point in terms of the other variables in this problem. Justify why this expression is independent of some variables and why it is dependent on the others.

2. 12 points; suggested time: 25–30 minutes

A physics teacher is riding an elevator while standing on a scale. She records the following values and makes the following notes as she rides the elevator upward from the ground floor to her desired floor. Her true weight is 600 N.

0–0.5 seconds	0.5–1.5 seconds	1.5–3 seconds	3–3.5 seconds	3.5–4 seconds
Scale reads 660 N	Scale gradually decreases from 660 N to 600 N	Scale reads a constant 600 N	Scale gradually decreases from 600 N to 480 N	Scale gradually increases from 480 N to 600 N

(a) Based on this information, sketch out a corresponding acceleration versus time graph for her trip.

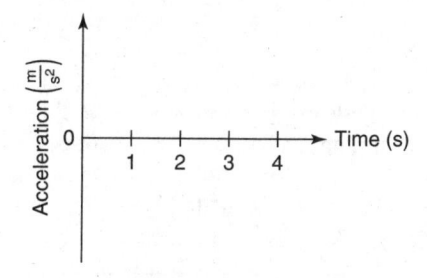

(b) Knowing that she both begins and ends her trip with zero velocity, sketch a corresponding velocity versus time sketch for her trip.

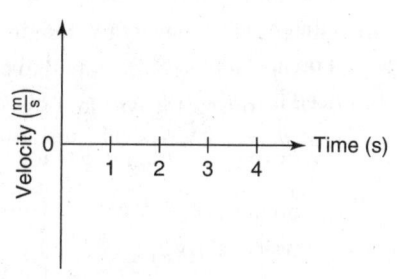

(c) Draw shaded rectangles to complete the bar charts for the gravitational potential energy U and the translational kinetic energy K when the physics teacher is halfway up and again when she is at her highest point. Her kinetic energy at 0.5 seconds and her potential energy at the highest point are completed for you. Fill in the missing four bar chart elements.
- Represent any energy that is equal to zero with a distinct line on the zero-energy line.
- The relative height of each shaded region should reflect the magnitude of the respective energy consistent with the scale shown.

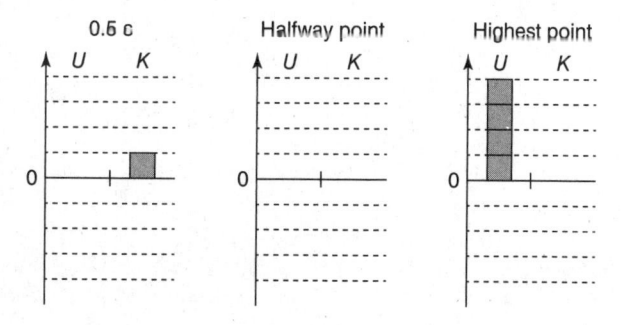

(d) For each of the following, determine whether the quantity is negative, zero, or positive. In each case, justify your answer.
   (i) What is the net work done on the physics teacher over the trip?
   (ii) What is the work done by gravity on the physics teacher over the trip?
   (iii) What is the work done by the normal force of the floor of the elevator on the physics teacher over the trip?

3. 10 points; suggested time: 25–30 minutes

A group of students has been asked to determine the spring constant k of the following setup. The students have a known mass and a very smooth tabletop. The only additional equipment they have are metersticks and rulers.

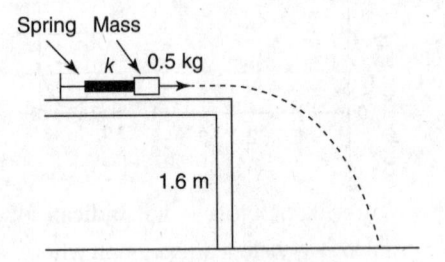

(a) (i) Describe an experimental procedure to determine the spring constant k. You may wish to label the diagram above further to help in your description. Indicate what measurements you would take and how you would take them. Include any steps necessary to reduce experimental uncertainty.
   (ii) Describe how the data collected in part (a) could be plotted to create a linear graph and how that graph would be analyzed to determine the spring constant k of the spring.

(b) A group of students do a different experiment in which they hang a single mass of 0.5 kg from a vertical spring and generate the following graph of its motion.

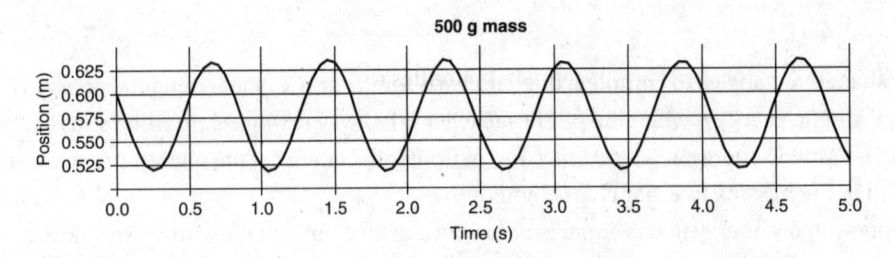

(i) From this data, describe how the students can determine the spring constant k.

(ii) The students then perform this experiment two more times with different masses on the same spring, resulting in the following graphs for a 250 g mass and a 1 kg mass.

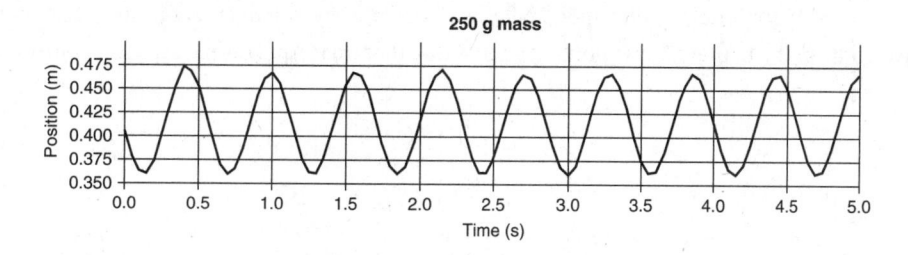

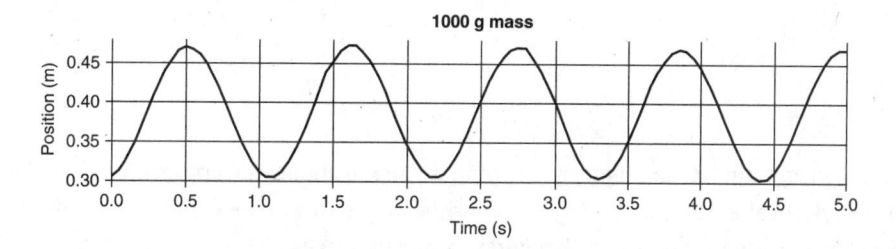

250 g mass

1000 g mass

Using the three datasets, extract three data points to plot. Indicate which measured or calculated quantity could be plotted on the horizontal axis to yield a linear graph whose slope can be used to calculate an experimental value for the spring constant. Fill in the chart below with the three relevant data points and the values to be graphed.

	y-axis	x-axis
Value to be graphed:		
Data point 1:		
Data point 2:		
Data point 3:		

(iii) On the grid below, plot the appropriate quantities to determine the spring constant. Clearly scale and label all axes, including units, as appropriate.

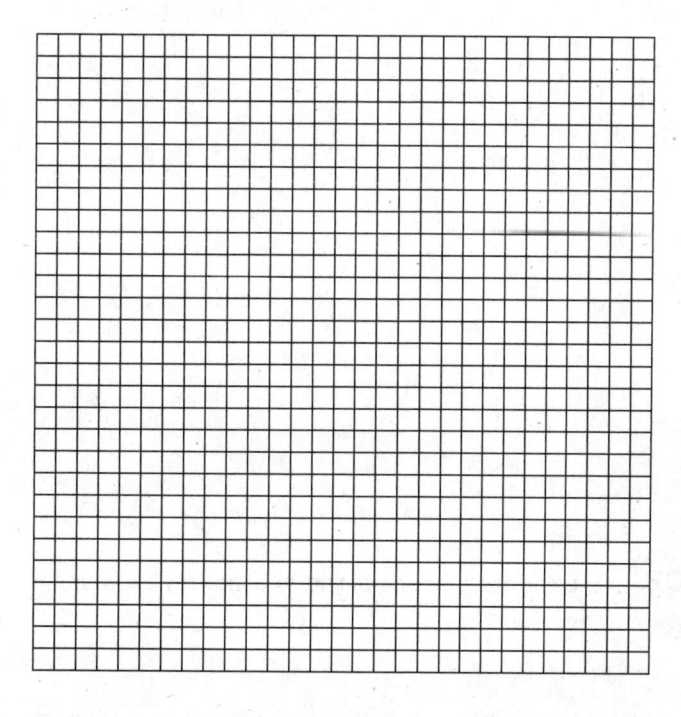

4. 8 points; suggested time: 15–20 minutes

A mass m is initially resting at the center of a rotating disk with rotational inertia $(1/2)MR^2$. The entire system is freely rotating with an initial angular speed of ω_0. As time goes by, the mass is carefully moved at a constant speed outward from the center. This is done in such a way that no torque on the system is introduced.

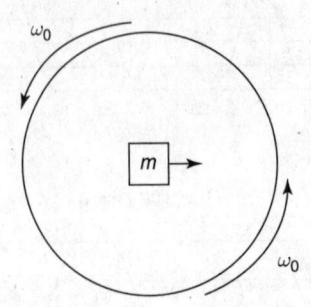

(a) If the mass m is about half the mass of the spinning disk, estimate the approximate angular speed of the two-mass system after the mass m has been pushed all the way to the edge in terms of ω_0. Justify your estimate using qualitative reasoning beyond referencing equations.
(b) How can one push the small mass m such that no torque is induced?
(c) Starting with the conservation of angular momentum, derive an equation for the angular velocity of the system as a function of r (the distance from the center at which mass m is currently located).
(d) Does your equation in part (c) agree with your qualitative reasoning in part (a)? Justify why or why not.

Answers Explained

Section I: Multiple-Choice

1. **(B)** Work $= fd \cos 60° = (10 \text{ N})(2.5)(0.5) = 12.5$ J

2. **(C)**
$$\text{Power} = \text{work/time} = (fd \cos 120°)/T = (mg \cos 120°)(D/T)$$
$$= (3)(10)(-0.5)(10 \text{ m/s}) = -150 \text{ W}$$

3. **(C)**
$$W = \Delta KE$$
$$\mu ND \cos 180° = 0 - \frac{1}{2}mv^2$$
$$(\mu mgD)(-1) = -\frac{1}{2}mv^2$$
$$D = v^2/2\mu g = 100/(2 \times 0.2 \times 10) = 25 \text{ m}$$

4. **(D)** Conservation of energy:
$$KE = U_s = \frac{1}{2}kx^2$$
Since we need x^2 to double, increase x by $(2)^{1/2} = \sqrt{2}$.

5. **(D)** The 30° and 60° angles will have equal ranges but with the roles of V_x and V_y reversed. Therefore, the projectile launched at a 30° angle will not go as high. Note that mass does not enter into projectile motion problems.

6. **(B)** First determine an expression for the angular acceleration:
$$\alpha = \Delta\omega/t = (\omega_2 - \omega_1)/0.5 = 2(\omega_2 - \omega_1)$$
Now use this in the expression for net torque:
$$\tau = I\alpha = I(2(\omega_2 - \omega_1))$$
Finally, substitute in the given expression for the rotational inertia:
$$\tau = (1/12)ML^2(2(\omega_2 - \omega_1)) = (\omega_2 - \omega_1)ML^2/6$$

7. **(C)**
$$\frac{1}{2}kx^2 = mgh$$
$h = kx^2/2mg$ above the starting point. Since the release point is a distance x above the starting point, we must subtract x from the answer.

8. **(D)** $F = kx$. So when graphing f versus x, k will be the slope. The extension of the spring is x when F is the force applied to the spring.

9. **(A)**
$$\vec{F}\Delta t = \Delta\vec{p}$$
$$(\vec{F})(15 \text{ s}) = 0 - (2{,}500)(30) = -75{,}000 \text{ kg} \cdot \text{m/s}$$
$$\vec{F} = -5{,}000 \text{ N}$$

The negative sign indicates an opposing force. Note that weight is not needed in this calculation, so the value of g is irrelevant.

10. **(C)** Momentum is always conserved. However, kinetic energy is lost unless the collision is elastic, in which case the kinetic energy is also conserved:

$$\vec{p}_i = 1(+6) + 2(+3) = 12 \text{ kg m/s} = \vec{p}_f = 3\vec{v}_f$$
$$\vec{v}_f = 4 \text{ m/s}$$

Initial *KE*:

$$\left(\frac{1}{2}\right)(1)(6)^2 + \left(\frac{1}{2}\right)(2)(3)^2 = 27 \text{ J}$$

Final *KE*:

$$\left(\frac{1}{2}\right)(3)(4)^2 = 24 \text{ J}$$

3 joules are lost.

11. **(D)** Impulse $= \Delta p = p_f - p_i = 0.2(-1) - (0.2)(3) = -0.8$ kg·m/s

12. **(B)**

$$\frac{1}{2}kA^2 = \text{total energy when } x = 4 \text{ (all } PE, \text{ no } KE)$$
$$\frac{1}{2}mv_{\text{max}}^2 = \text{total energy when } x = 0 \text{ (all } KE, \text{ no } PE)$$

Conservation of energy:

$$\frac{1}{2}kA^2 = \frac{1}{2}mv_{\text{max}}^2$$
$$v_{\text{max}} = (k/m)^{1/2} A$$

13. **(B)** In one cycle, the mass travels 4 amplitudes:

$$A = 0.02 \text{ m}$$
$$\text{Energy} = \frac{1}{2}kA^2 = \frac{1}{2}(50)(0.02)^2 = 0.01 \text{ J}$$

14. **(D)** Orbital velocity:

$$F_g = \text{centripetal force} = mv^2/r$$
$$GMm/r^2 = mv^2/r$$
$$v = (GM/r)^{1/2}$$

Escape velocity:

$$E_{\text{total}} > 0$$
$$KE + PE > 0$$
$$\frac{1}{2}mv^2 - GMm/r > 0 \text{ (using universal gravitational } PE)$$
$$v > (2GM/r)^{1/2}$$

15. **(B)** The gravitational force between two masses has a $1/r^2$ relationship (where *r* is the distance from center to center). Graph B shows the correct inverse relationship.

16. **(D)** Net work done to an object results in a change in kinetic energy. In this case the book has no kinetic energy at the beginning and at the end of its trip:

$$W_{\text{net}} = \Delta K$$

17. **(A)** The brick is experiencing three forces: a vertical normal and weight force that cancel out as well as a horizontal force of friction opposing the motion. This force is unopposed and is the cause of the deceleration of the brick. The inertia of the brick is provided by the fact that it is a mass in motion and would continue to move at constant velocity if there were no friction present. The pushing force is no longer active on the brick and plays no role in the analysis.

18. **(C)** At maximum height, all speed is from V_x:

$$V_x = 200 \cos 30° = 173 \text{ m/s}$$

19. **(A)** Vertical forces must cancel:

$$T \cos \theta - mg = 0$$
$$T = mg/\cos \theta$$

20. **(D)** $v = p/m = (2 \times 1 + 3)/2 = 2.5$ m/s

21. **(B)** Since $\vec{F}\Delta t = \Delta \vec{p}$, a net force will change the linear momentum. However, torque also involves the lever arm ($\tau = RF\sin\theta$). So despite having a nonzero net force, the net torque might still be zero.

22. **(C)** Circular motion:

$$F_{net} = mv^2/r$$
$$mg - N = mv^2/r$$

N is the force the car exerts on the driver:

$$mg - mv^2/r = 500 - 50(20^2)/50$$
$$= 500 - 400 = 100 \text{ N}$$

Note that the correct answer can be found by simply knowing that the net force must be down and that the car must be pushing upward on the passenger.

23. **(D)** Take the center of mass to be 1 meter from the pivot:

$$\text{Net torque} = -mg(1)\sin 90° + 200(2)\sin 30°$$
$$= 98 + 200 = 102 \text{ N} \cdot \text{m}$$

24. **(B)** Acceleration is the slope on velocity versus time graphs. Since there are four separate intervals of straight lines, there are four intervals of constant acceleration: positive for the first 10 seconds, zero for the subsequent 5 seconds, a negative acceleration from the 15th to the 40th second, and positive for the last 15 seconds.

25. **(D)** To find the average acceleration for any interval on a velocity vs. time graph, connect the beginning of the interval to the end of the interval and find the slope of that line. In this case, the graph begins and ends on the axis and therefore the connecting line is horizontal: slope = 0.

26. **(B)** Displacement is area under the curve: $+1{,}050 - 500 = 550$ m

27. **(D)** What is the instantaneous acceleration is the slope at that point: $-100/25 = -4$ m/s/s

28. **(C)**

$$\text{Torque} = I\alpha$$
$$\alpha = \text{angular acceleration} = \Delta\omega/\Delta t$$

Our only data provided are that the angular acceleration must be larger for the 75 RPM record and that the rotational inertia has not changed (same disk). Therefore, the torque supplied must be larger.

29. **(A)** Remember conservation of energy. Since both start at the same height (same potential energy) with the same kinetic energy, they will both hit the ground with the same joules of energy (all kinetic energy). The same kinetic energy means the same speed since their masses are the same. Note that this does not imply that both components of velocity are the same. They are not. This implies that only the magnitudes of the final velocity vectors are the same.

30. **(C)** The vertically projected sphere will spend much more time in the air as it goes much higher. Without knowing the exact speed of the launch, it is not possible to say by what factor the time in flight is extended.

31. **(A)** The change in gravitational potential energy is the same for both. Therefore, the work done by gravity is the same.

32. **(C)** First use the definition of density to find an expression for the mass of the cube:

$$mass = \rho_c V$$

Since this is not one of the answers, find an expression for equilibrium using Archimedes' principle of buoyancy:

$$F_B = W$$
$$\rho_L(V/2)g = (mass)g$$

Solve for mass:

$$mass = \rho_L(V/2)$$

33. **(D)** The net force is the difference between the buoyancy force and the weight of the cube:

$$F_{net} = F_B - W$$
$$F_{net} = \rho_L V g - \rho_c V g$$
$$F_{net} = (\rho_L - \rho_c)Vg$$

Since this is the net force, to find the resulting acceleration, divide by the mass of the cube:

$$mass = \rho_c V$$
$$a = (\rho_L - \rho_c)Vg/\rho_c V$$
$$a = (\rho_L - \rho_c)g/\rho_c$$

34. **(B)** The angular displacement is the area under the given graph:

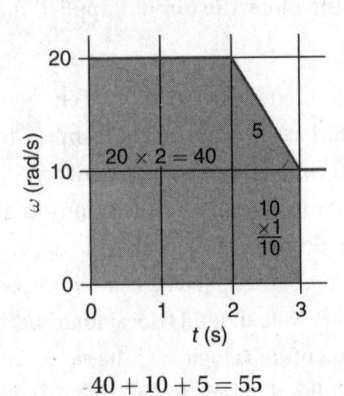

$$40 + 10 + 5 = 55$$

35. **(A)** The change in angular momentum is the torque multiplied by the time:

$$\Delta L = \tau \Delta t$$

Since each time interval is only 1 second:

$$\Delta L = \tau(1) = I\alpha$$

where α is the slope of the line in each interval.

	0–1 seconds	1–2 seconds	2–3 seconds
Slope	0	0	$-10I$

36. **(D)** In a horizontal plane, the tension is the only centripetal force. However, in a vertical plane, the tension and the force due to gravity are both along the radial line. At the top, gravity is working downward along with the tension so $F_{net} = T + mg$ whereas, at the bottom, gravity is still downward while the tension is still directed inward $F_{net} = T - mg$. In all cases $F_{net} = MV^2/R$.

37. **(C)** From her first experiment:

$$F_{net} = ma$$
$$KX = ma \quad (1)$$

From her second experiment:

$$T = 2\pi(m/K)^{1/2} \quad (2)$$

The only two measurement values she has are T and X, so eliminating K and m from these two equations and solving for a:

$$a = KX/m \quad (1)$$

and

$$K = m(2\pi/T)^2 \quad (2)$$

Combining:

$$a = (2\pi/T)^2 X$$

38. **(D)** The elevator must be traveling at constant velocity to insure the normal force (reading on the scale) is the same as her true weight. However, this velocity value can be any number: positive, zero, or negative.

39. **(C)** Remember that $g = Gm/r^2$. Half the mass means half the g. Half the r means $4 \times g$. Collect the changes:

$$\frac{1}{2} \times 4 = 2$$

40. **(C)** Pressure equals density times height times g. The tall tube of water will have the most height and thus the greatest pressure at its bottom surface.

Section II: Free-Response

1. (a) 1 point for showing the friction force is up the hill and begins at the point of contact between the ramp and the disk.

   1 point for indicating this produces a counterclockwise torque.

   1 point for stating that normal and gravitational forces produce no torque.

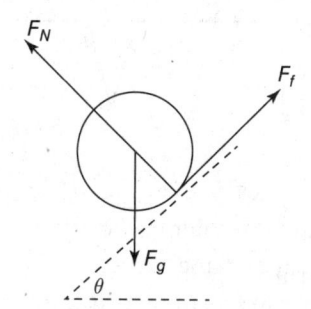

   (b) 1 point

   $$\text{Net torque} = F_f R \sin 90° = F_f R$$

   Since the rotation is counterclockwise, this torque will decrease the angular speed.

   (c) 1 point for the general expression of energy.

   $$Mgh + (1/2)Mv^2 + (1/2)I\omega^2$$

   1 point for putting in the expression for the rotational inertia and the rolling without slipping constraint for ω:

   $$Mgh + (1/2)Mv^2 + (1/2)((2/5)MR^2)(v/r)^2$$
   $$Mgh + (1/2)Mv^2 + (1/5)Mv^2$$
   $$Mgh + (7/10)Mv^2$$

   1 point for setting this expression equal to the initial kinetic energy, $(7/10)Mv_0^2$, before the mass goes uphill.

   $$Mgh + (7/10)Mv^2 = (7/10)Mv_0^2$$
   $$h = (7/10g)(v_0^2 - v^2)$$

   (d) 1 point for stating $v = 0$ at the highest point.

   1 point for a final expression for h.

   $$h = 7v_0^2/10g$$

   1 point. Since all rotational motion has ceased at the highest point, this expression for height is dependent on only the initial speed, which determines both rotational and translational kinetic energies. The actual angle is irrelevant because height is the only determinant of potential energy and no work is done by nonconservative forces in this problem. (Note that no work is done by friction because no displacement is occurring at the point of contact.)

2. (a) 0.5 point for a horizontal line for the first 0.5 seconds.

   0.5 point for a linear decrease to 0 over the next second.

   0.5 point for a horizontal line from 1.5 to 3 seconds.

   0.5 point for a negative peak at -2 m/s^2.

Acceleration values are determined using the following formula.

$$a = (F_n - F_g)/60 \text{ kg}$$

(b) 0.5 point for a straight line that increases to 0.5 m/s for 0.5 seconds.
0.5 point for a curve that increases to 1 m/s during the next second.
0.5 point for a constant 1 m/s velocity until the 3 s mark.
0.5 point for alternating curves down to zero during the last second.

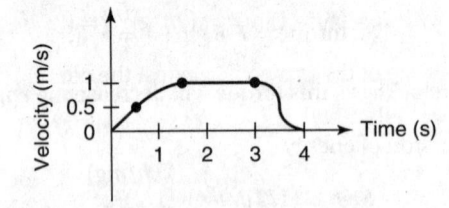

(c) 1 point for having 4 times as much kinetic energy at the halfway point as at the 0.5 s mark (doubling velocity leads to a quadrupling of kinetic energy).
1 point for half the potential energy at the halfway point and 0 kinetic energy at the top.
1 point for a small amount of potential energy at the halfway point. This small amount is approximately 1/6 of a block and is based off of the area under the curves in part (b).

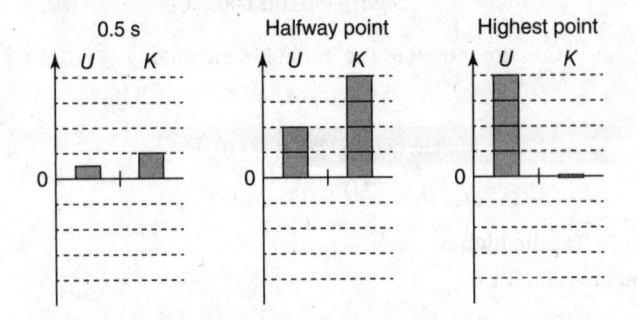

(d) 1 point each.
   (i) Net work is 0 since there is no net change in kinetic energy.
   (ii) Work done by gravity is negative since the displacement is upward and the force of gravity is downward.
   (iii) Work done by the normal force is positive since the normal force is upward and the displacement is upward.

3. (a) (i) 1 point. For various measured compressions of the spring (x), measure the horizontal range for the mass (R). Range should be measured along the floor from beneath the edge of the table to where the mass first hits the ground. Multiple trials for each compression x should be taken so that the average range of values can be determined.

(ii) 1 point. Set the initial potential energy of the spring equal to the kinetic energy off of the table.

$$(1/2)kx^2 = (1/2)mv^2$$
$$v = (k/m)^{1/2}x$$

This velocity is the horizontal projectile's velocity. The time in flight is found from the height of the table.

$$h = (1/2)gt^2$$
$$t = (2h/g)^{1/2}$$

Multiplying the horizontal velocity and time in flight gives us the range of the particle.

$$R = vt = (k/m)^{1/2}x(2h/g)^{1/2} = (2kh/mg)^{1/2}x$$

1 point. By plotting the range of the projectile against the compression of the spring, the spring constant can be determined by the slope.

$$\text{Slope} = (2kh/mg)^{1/2}$$

So for a known mass and height of table, the spring constant k can be determined from this slope.

(b) (i) 1 point. By measuring the period of the oscillation and using the known mass of the oscillation, the spring constant can be determined from the following equation.

$$T = 2\pi(m/k)^{1/2}$$

(ii) 1 point for identifying which values belong on the x-axis and which belong on the y-axis and for including proper units.

1 point for including correct values.

	y-axis	x-axis
Value to be graphed:	(Period)^2 (s^2)	Mass (kg)
Data point 1:	$(4/7)^2 = 0.33$	0.25
Data point 2:	$(0.9)^2 = 0.81$	0.50
Data point 3:	$(1.1)^2 = 1.2$	1.0

(iii) 1 point for including the labels and units on the graph.
1 point for the line of best fit.
1 point for determining the slope.

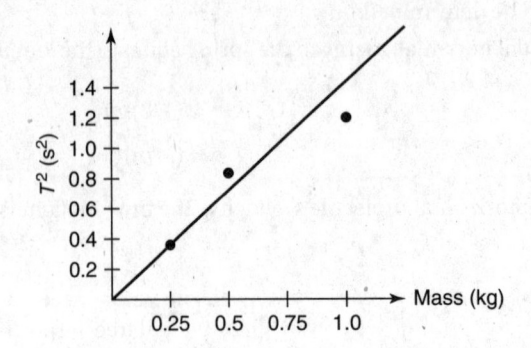

1 point.

$$\text{Slope} = (1/0.75)$$
$$k = 4\pi^2(0.75) = 30. \text{ N/m}$$

4. (a) 1 point. Pushing the mass out to the edge will result in an increase in the rotational inertia of the system of mr^2 or $(M/2)r^2$, which is same as the original rotational inertia of the disk alone.
1 point. Since this is effectively doubling the rotational inertia, the angular speed should be reduced to 1/2 of its original value.

(b) 1 point. If the force is applied radially outward the entire time, the lever arm for the torque will be parallel to the force and thus no torque will result.

(c) 1 point for the initial angular momentum.

$$L = I\omega_0 = (1/2)Mr^2\omega_0$$

1 point for stating that as the small mass is pushed outward, it will contribute to the rotational inertia.

$$L = I\omega + (mr^2)\omega = (1/2)Mr^2\omega + (mr^2)\omega = r^2\omega(M/2 + m)$$

1 point for first setting these expressions equal to each other (conservation of angular momentum)

$$(1/2)Mr^2\omega_0 = r^2\omega(M/2 + m)$$

and then solving for angular velocity.

$$\omega = \omega_0(M/(M + 2m))$$

(d) 1 point for substituting in $m = M/2$.

$$\omega = \omega_0(M/(M + 2(M/2))) = \omega_0(M/(2M)) = \omega_0/2$$

1 point for stating that the equation in part (c) agrees with the answer in part (a).

PRACTICE TEST 2

Test Analysis

Practice Test 2

Section I: Multiple-Choice

Note that the questions requiring two answers are to be graded as completely correct (1 point) or incorrect (0 points).

Number correct (out of 40) = $\dfrac{}{\text{Multiple-Choice Score}}$

Section II: Free-Response

Partial credit is awarded for any correct responses within an individual free-response question.

Question 1 = $\dfrac{}{\text{(out of 10)}}$

Question 2 = $\dfrac{}{\text{(out of 12)}}$

Question 3 = $\dfrac{}{\text{(out of 10)}}$

Question 4 = $\dfrac{}{\text{(out of 8)}}$

Total = $\dfrac{}{\text{(out of 80)}}$

Final Score

$\dfrac{}{\text{Multiple-Choice Score}} + \dfrac{}{\text{Free-Response Score}} = \dfrac{}{\text{Total (out of 80)}}$

Final Score Range*

Final Score Range	AP Score
60–80	5
44–59	4
32–43	3
20–31	2
0–19	1

*Note: The guidelines above are based on the released scores for past AP Physics 1 exams. Actual score ranges vary from year to year and are determined by the College Board each year. Thus, the ranges shown are approximate.

Appendix

Table of Information for AP Physics 1

Useful Constants

Acceleration due to gravity at Earth's surface	$g = 9.8 \text{ m/s}^2 = 9.8 \text{ N/kg}$
Universal gravitational constant	$G = 6.67 \times 10^{-11} \text{ m}^3/(\text{kg} \cdot \text{s}^2)$
	$= 6.67 \times 10^{-11} \text{ Nm}^2/\text{kg}^2$
1 atmosphere of pressure	$1 \text{ atm} = 10^5 \text{ Pa} = 10^5 \text{ N/m}^2$

Unit Symbols

Meter, m	Second, s	Newton, N	Watt, W
Kilogram, kg	Hertz, Hz	Joule, J	Pascal, Pa

Prefixes

Factor	Prefix	Symbol
10^{12}	Tera-	T
10^9	Giga-	G
10^6	Mega-	M
10^3	Kilo-	k
10^{-2}	Centi-	c
10^{-3}	Milli-	m
10^{-6}	Micro-	μ
10^{-9}	Nano-	n
10^{-12}	Pico-	p

Formula Sheet for AP Physics 1

Mechanics

$v = v_0 + at$

$x = x_0 + v_0 t + \frac{1}{2}at^2$

$v^2 = v_0^2 + 2a(x - x_0)$

$\sum \vec{F} = \vec{F}_{net} = m\vec{a}$

$F_{fric} \leq \mu N$

$a_c = \frac{v^2}{r}$

$\tau = rF\sin\theta$

$\vec{p} = m\vec{v}$

$\vec{J} = \vec{F}\Delta t = \Delta \vec{p}$

$K = \frac{1}{2}mv^2$

$\Delta U_g = mg\Delta y$

$\Delta E = W = Fd\cos\theta$

$P_{avg} = \frac{\Delta E}{\Delta t}$

$\theta = \theta_0 + \omega_0 t + \frac{1}{2}\alpha t^2$

$\omega = \omega_0 + \alpha t$

$x = A\cos(2\pi ft)$

$I = \sum m_i r_i^2$

$I' = I_{cm} + Md^2$

$\vec{v}_{cm} = \frac{\sum \vec{p}_i}{\sum m_i} = \frac{\sum (m_i \vec{v}_i)}{\sum m_i}$

$\vec{x}_{cm} = \frac{\sum m_i \vec{x}_i}{\sum m_i}$

$T = rF\sin\theta$

$L = I\omega = rmv\sin(\theta)$

$\Delta L = \tau \Delta t$

$K = \frac{1}{2}I\omega^2$

$\rho = m/V$

$\vec{F}_s = -k\vec{x}$

$U_s = \frac{1}{2}kx^2$

$T_s = 2\pi\sqrt{\frac{m}{k}}$

$T_p = 2\pi\sqrt{\frac{\ell}{g}}$

$T = \frac{1}{f} = \frac{2\pi}{\omega}$

$F_G = \frac{Gm_1 m_2}{r^2}$

$U_G = -\frac{Gm_1 m_2}{r}$

a = acceleration
F = force
f = frequency
h = height
I = rotational inertia
K = kinetic energy
k = spring constant
L = angular momentum
ℓ = length
m = mass
N = normal force
P = power
p = momentum
r = radius or distance
T = period

t = time
U = potential energy
V = volume
v = velocity or speed
ω = angular speed
W = work done on a system
x = position
α = angular acceleration
θ = angle
ρ = density
τ = torque
μ = coefficient of friction

Fluids

$P = \frac{F_\perp}{A}$

$P = P_0 + \rho g h$

$P_{gauge} = \rho g h$

$F_b = \rho V g$

$A_1 v_1 = A_2 v_2$

$P_1 + \rho g y_1 + \frac{1}{2}\rho v_1^2 = P_2 + \rho g y_2 + \frac{1}{2}\rho v_2^2$

P = pressure
A = cross-sectional area
ρ = density

Geometry and Trigonometry

Rectangle
$A = bh$

Triangle
$A = \frac{1}{2}bh$

Circle
$A = \pi r^2$
$C = 2\pi r$

Parallelepiped
$V = \ell w h$

Cylinder
$V = \pi r^2 \ell$
$S = 2\pi r \ell + 2\pi r^2$

Sphere
$V = \frac{4}{3}\pi r^3$
$S = 4\pi r^2$

Right Triangle
$a^2 + b^2 = c^2$
$\sin\theta = \frac{a}{c}$
$\cos\theta = \frac{b}{c}$
$\tan\theta = \frac{a}{b}$

A = area
C = circumference
V = volume
S = surface area
b = base
h = height
ℓ = length
w = width
r = radius

Glossary

acceleration A vector quantity representing the time rate of change of velocity.

action A force applied to an object that leads to an equal but opposite reaction; the product of total energy and time in units of joules times seconds; the product of momentum and position, especially in the case of the Heisenberg uncertainty principle.

alternating current An electric current that changes its direction and magnitude according to a regular frequency.

ammeter A device that, when placed in series, measures the current in an electric circuit; a galvanometer with a low-resistance coil placed across it.

ampere (A) The SI unit of electric current, equal to 1 C/s.

amplitude The maximum displacement of an oscillating particle, medium, or field relative to its rest position.

angstrom (Å) A unit of distance measurement equal to 1×10^{-10} m.

antinodal line A region of maximum displacement in a medium where waves are interacting with each other.

average speed A scalar quantity equal to the ratio of the total distance to the total elapsed time.

battery A combination of two or more electric cells.

beats The interference caused by two sets of sound waves with only a slight difference in frequency.

Cartesian coordinate system A set of two or three mutually perpendicular reference lines, called axes and usually designated as x, y, and z, that are used to define the location of an object in a frame of reference; a coordinate system named for French scientist Rene Descartes.

center of curvature A point that is equidistant from all other points on the surface of a spherical mirror; a point equal to twice the focal length of a spherical mirror.

center of mass The weighted mean distribution point where all the mass of an object can be considered to be located; the point at which, if a single force is applied, translational motion will result.

centripetal acceleration The acceleration of mass moving in a circular path directed radially inward toward the center of the circular path.

centripetal force The deflecting force, directed radially inward toward a given point, that causes an object to follow a circular path.

coefficient of friction The ratio of the force of friction to the normal force when one surface is sliding (or attempting to slide) over another surface.

coherent Referring to a set of waves that have the same wavelength, frequency, and phase.

component One of two mutually perpendicular vectors that lie along the principal axes in a coordinate system and can be combined to form a given resultant vector.

concurrent forces Two or more forces that act at the same point and at the same time.

conductor A substance, usually metallic, that allows the relatively easy flow of electric charges.

conservation of energy A principle of physics that states that the total energy of an isolated system remains the same during all interactions within the system.

conservation of momentum A principle of physics that states that, in the absence of any external forces, the total momentum of an isolated system remains the same.

conservative force A force such that any work done by this force can be recovered without any loss; a force whose work is independent of the path taken.

constructive interference   The additive result of two or more waves interacting with the same phase relationship as they move through a medium.

coordinate system   A set of reference lines, not necessarily perpendicular, used to locate the position of an object within a frame of reference by applying the rules of analytic geometry.

coulomb (C)   The SI unit of electrical charge, defined as the amount of charge 1 A of current contains each second.

Coulomb's law   The electrostatic force between two point charges is directly proportional to the product of the charges and inversely proportional to the square of the distance separating them. Named for French physicist Charles-Augustin de Coulomb.

current   A scalar quantity that measures the amount of charge passing a given point in an electric circuit each second.

cycle   One complete sequence of periodic events or oscillations.

damping   The continuous decrease in the amplitude of mechanical oscillations due to a dissipative force.

deflecting force   Any force that acts to change the direction of motion of an object.

derived unit   Any combination of fundamental physical units.

destructive interference   The result produced by the interaction of two or more waves with opposite phase relationships as they move through a medium.

diffraction   The ability of waves to pass around obstacles or squeeze through small openings.

direct current   Electric current that is moving in one direction only around an electric circuit.

displacement   A vector quantity that determines the change in position of an object by measuring the straight-line distance and direction from the starting point to the ending point.

dissipative force   Any force, such as friction, that removes kinetic energy from a moving object; a nonconservative force.

distance   A scalar quantity that measures the total length of the path taken by a moving object.

Doppler effect   The apparent change in the wavelength or frequency of a wave as the source of the wave moves relative to an observer. Named for Austrian physicist Christian Doppler.

dynamics   The branch of mechanics that studies the effects of forces on objects.

elastic collision   A collision between two objects in which there is a rebounding and no loss of kinetic energy occurs.

elastic potential energy   The energy stored in a spring when work is done to stretch or compress it.

electrical ground   The passing of charges to or from Earth to establish a potential difference between two points.

electric cell   A chemical device for generating electricity.

electric circuit   A closed conducting loop consisting of a source of potential difference, conducting wires, and other devices that operate on electricity.

electromotive force (emf)   The potential difference caused by the conversion of different forms of energy into electrical energy; the energy per unit charge.

electron   A negatively charged particle that orbits a nucleus in an atom; the fundamental carrier of negative electric charge.

electroscope   A device for detecting the presence of static charges on an object.

elementary charge   The fundamental amount of charge of an electron.

emf   See **electromotive force**.

energy   A scalar quantity representing the capacity to work.

equilibrant   The force equal in magnitude and opposite in direction to the resultant of two or more forces that brings a system into equilibrium.

equilibrium   The balancing of all external forces acting on a mass; the result of a zero vector sum of all forces acting on an object.

escape velocity   The velocity attained by an object such that, if coasting, the object would not be pulled back toward the planet from which it came.

field   A region characterized by the presence of a force on a test body like a unit mass in a gravitational field or a unit charge in an electric field.

force   A vector quantity that corresponds to any push or pull due to an interaction of matter that changes the motion of an object.

force constant   See **spring constant**.

forced vibration   A vibration caused by the application of an external force.

frame of reference   A point of view consisting of a coordinate system in which observations are made.

free-body diagram   A diagram that illustrates all of the forces acting on a mass at any given time.

frequency   The number of completed periodic cycles per second in an oscillation or wave motion.

friction   A force that opposes the motion of an object as it slides over another surface.

fundamental unit   An arbitrary scale of measurement assigned to certain physical quantities, such as length, time, mass, and charge, that are considered to be the basis for all other measurements. In the SI system, the fundamental units used in physics are the meter, kilogram, second, ampere, kelvin, and mole.

galvanometer   A device used to detect the presence of small electric currents when connected in series in a circuit.

gravitation   The mutual force of attraction between two uncharged masses.

gravity   Another name for gravitation or the gravitational force; the tendency of objects to fall to Earth.

hertz (Hz)   The SI unit of frequency, equal to $1\ s^{-1}$.

Hooke's law   The stress applied to an elastic material is directly proportional to the strain produced. Named for English scientist Robert Hooke.

impulse   A vector quantity equal to the product of the average force applied to a mass and the time interval in which the force acts; the area under a force versus time graph.

inelastic collision   A collision in which two masses interact and stick together, leading to an apparent loss of kinetic energy.

inertia   The property of matter that resists the action of applied force trying to change the motion of an object.

inertial frame of reference   A frame of reference in which the law of inertia holds; a frame of reference moving with constant velocity relative to Earth.

instantaneous velocity   The slope of a tangent line to a point in a displacement versus time graph.

insulator   A substance that is a poor conductor of electricity because of the absence of free electrons.

interference   The interaction of two or more waves, producing an enhanced or a diminished amplitude at the point of interaction; the superposition of one wave on another.

interference pattern   The pattern produced by the constructive and destructive interference of waves generated by two point sources.

isolated system   A combination of two or more interacting objects that are not being acted upon by external force.

joule (J)   The SI unit of work, equal to $1\ N \cdot m$; the SI unit of mechanical energy, equal to $1\ kg \cdot m^2/s^2$.

junction   The point in an electric circuit where a parallel connection branches off.

Kepler's first law   The orbital paths of all planets are elliptical. Named for German astronomer Johannes Kepler.

Kepler's second law   A line from the Sun to a planet sweeps out equal areas in equal time.

Kepler's third law   The ratio of the cube of the mean radius to the Sun to the square of the period is a constant for all planets orbiting the Sun.

kilogram (kg)   The SI unit of mass.

kilojoule (kJ)   A unit representing 1,000 J.

kinematics   In mechanics, the study of how objects move.

kinetic energy   The energy possessed by a mass because of its motion relative to a frame of reference.

kinetic friction   The friction induced by sliding one surface over another.

Kirchhoff's junction rule   The algebraic sum of all currents at a junction in a circuit equals zero. Named for German physicist Gustav Kirchhoff.

Kirchhoff's loop rule The algebraic sum of all potential drops around any closed loop in a circuit equals zero.

law of inertia See **Newton's first law of motion**.

longitudinal wave A wave in which the oscillating particles vibrate in a direction parallel to the direction of propagation.

mass The property of matter used to represent the inertia of an object; the ratio of the net force applied to an object and its subsequent acceleration as specified by Newton's second law of motion.

mean radius For a planet, the average distance to the Sun.

meter (m) The SI unit of length.

moment arm distance A line drawn from a pivot point.

momentum A vector quantity equal to the product of the mass and the velocity of a moving object.

natural frequency The frequency with which an elastic body will vibrate if disturbed.

net force The resultant force acting on a mass.

newton (N) The SI unit of force, equal to 1 kg · m/s^2.

Newton's first law of motion An object at rest tends to remain at rest, and an object in motion tends to remain in motion at a constant velocity, unless acted upon by an external force; the law of inertia. Named for British mathematician/physicist Sir Isaac Newton.

Newton's second law of motion The acceleration of a body is directly proportional to the applied net force; $\vec{F} = m\vec{a}$.

Newton's third law of motion For every action, there is an equal, but opposite, reaction.

nodal line A line of minimum displacement when two or more waves interfere.

nonconservative force A force, such as friction, that decreases the amount of kinetic energy after work is done by or against the force.

normal A line perpendicular to a surface.

normal force A force that is directed perpendicularly to a surface when two objects are in contact.

ohm (Ω) The SI unit of electrical resistance, equal to 1 V/A.

Ohm's law In a circuit at constant temperature, the ratio of the potential difference to the current is a constant (called the resistance). Named for German physicist Georg Ohm.

parallel circuit A circuit in which two or more devices are connected across the same potential difference and provide an alternative path for charge flow.

period The time, in seconds, to complete one cycle of repetitive oscillations or uniform circular motion.

phase The relative position of a point on a wave with respect to another point on the same wave.

polar coordinate system A coordinate system in which the location of a point is determined by a vector from the origin of a Cartesian coordinate system and the acute angle it makes with the positive horizontal axis.

potential difference The difference in work per unit charge between any two points in an electric field.

potential energy The energy in a system resulting from the relative position of two objects interacting via a conservative force.

power The ratio of the work done to the time needed to complete the work.

pressure The force per unit area.

pulse A single vibratory disturbance in an elastic medium.

resistance The ratio of the potential difference across a conductor to the current in the conductor.

resistivity A property of matter that measures the ability of that substance to act as a resistor.

resolution of forces The process by which a given force is decomposed into a pair of perpendicular forces.

resonance The production of sympathetic vibrations in a body, at its natural vibrating frequency, caused by another vibrating body.

rotational equilibrium The situation when the vector sum of all torques acting on a rotating mass equals zero.

rotational inertia The ability of a substance to resist the action of a torque; moment of inertia.

scalar A physical quantity, such as mass or speed, that is characterized by magnitude only.

second (s) The SI unit of time.
series circuit A circuit in which electrical devices are connected sequentially in a single conducting loop, allowing only one path for charge flow.
sliding friction A force that resists the sliding motion of one surface over another; kinetic friction.
speed A scalar quantity measuring the time rate of change of distance.
spring constant The ratio of the applied force and resultant displacement of a spring.
standing wave A stationary wave pattern formed in a medium when two sets of waves with equal wavelength and amplitude pass through each other, usually after a reflection.
static electricity Stationary electric charges.
static friction The force that prevents one object from sliding over another.
statics The study of the forces acting on an object that is at rest relative to a frame of reference.
superposition The ability of waves to pass through each other, interfere, and then continue on their way unimpeded.
torque The application, from a pivot, of a force at right angles to a designated line that tends to produce circular motion; the product of a force and a perpendicular moment arm distance.
total mechanical energy In a mechanical system, the sum of the kinetic and potential energies.
transverse wave A wave in which the vibrations of the medium or field are at right angles to the direction of propagation.
uniform circular motion Motion around a circle at a constant speed.
uniform motion Motion at a constant speed in a straight line.
unit An arbitrary scale assigned to a physical quantity for measurement and comparison.
universal law of gravitation The gravitational force between any two masses is directly proportional to the product of their masses, and inversely proportional to the square of the distance between them.
vector A physical quantity that is characterized by both magnitude and direction; a directed arrow drawn in a Cartesian coordinate system used to represent a quantity such as force, velocity, or displacement.
velocity A vector quantity representing the time rate of change of displacement.
volt (V) The SI unit of potential difference, equal to 1 J/C.
voltmeter A device used to measure the potential difference between two points in a circuit when connected in parallel; a galvanometer with a high-resistance coil placed in series with it.
watt (W) The SI unit of power, equal to 1 J/s.
wave A series of periodic disturbances in an elastic medium or field.
wavelength The distance between any two successive points in phase on a wave.
weight The force of gravity exerted on a mass at the surface of a planet.
work A scalar measure of the relative amount of change in mechanical energy; the product of the magnitude of the displacement of an object and the component of applied force in the same direction as the displacement; the area under a graph of force versus displacement.

Index

A
Acceleration, 46–52, 58, 77, 120
Action-reaction law, 69
Addition, vectors, 33–41
Amplitude, 150
Angular kinematics, 139–140
Angular momentum, 141–142
Angular motion, *versus* linear motion, 140
AP Physics
    objects *versus* systems, 12
    scope of exam, 1–2
Archimedes' principle, 163–164
Average velocity, 45

B
Bar, 161
Bernoulli's equation, 165–168
Buoyancy principle, 163–164
Buoyant force, 164

C
Cartesian system, 31–32
Cavendish, Henry, 108
Center of mass, 124–125
Central forces, 76
Centripetal acceleration, 58, 76, 88
Centripetal force, 8, 76, 85, 89
Circular motion, 57–59, 80, 151–152
Coefficient of friction, 77
Collision(s), 122–124
Component form, 33
Compressible fluids, 161
Conical pendulum, 76
Conservation
    of energy, 96–100
    of linear momentum, 121
Conservative forces, 96
Contact force, 67
Coordinate systems, 31
Cosine function, 35

D
Data collection strategies, use of, 9
Depth, 162–163
Direct relationship, 4
Direction, magnitude of vectors and, 33
Displacement vector, 45
Distance, 45–46

E
Elastic collisions, 122–123
Elastics, 96
Energy
    conservation, 96–100

friction, 77–80
gravitational, 109–110
kinetic, 94–95, 123
mechanical, 96
potential, 96, 109–110
power and, 94
rolling objects, 141
work, 91–93
work-energy theorem, 95
Equation of continuity, 165
Equilibrium, 67, 68, 138–139
Escape velocity, 110
Experimental design questions, 6
External force, 119

F
Fluid(s)
    Archimedes' principle, 163–164
    Bernoulli's equation, 165–168
    buoyancy principle, 163–164
    column of, 162
    escaping through small orifice, 166–167
    friction, 68
    in motion, 164–165
    moving horizontally, 167–168
    Pascal's principle, 161–162
    at rest, 166
    static, 161
    static pressure and depth, 162–163
Force(s)
    central, 76
    centripetal, 76, 80
    contact, 67
    internal and external, 119
    net, 119
    nonconservative, 98
    normal, 71, 72
    overview, 67–69
    types of, 68
Frames of reference, 31
Free fall, 51
Free response
    diagnostic test, 21
    tips for solving, 6–8
    vector practice, 41
Free-body diagrams, 72–75
Frequency, 58
Friction, 68, 74, 77–80
Fulcrum, 134

G
Galilei, Galileo, 69
Graphical analysis, of motion, 52–54
Graphs, tips for answering, 9–10

Gravitation, Newton's law of universal, 107–109
Gravitational acceleration, 51–52
Gravitational energy, 109–110
Gravity, 51–52, 68, 96

H
Halley, Edmund, 69
Harmonic motion. *See* Simple harmonic motion
Hooke's law, 75, 149
Horizontally launched projectiles, 55–56

I
Impacts
    change in momentum, 119–121
    collisions, 122–124
    forces, 119–121
    linear momentum, 121
Impulse, 120
Incompressible fluids, 161
Inelastic collisions, 123–124
Inertia, 69, 139
Inertial frame of reference, 31
Instantaneous velocity, 46, 52
Internal force, 119
Inverse relationship, 4
Inverse square law, 4, 108, 109, 113

J
Joules, 91

K
Kepler, Johannes, 69
Kepler's laws of planetary motion, 111–112
Kinematics
    acceleration, 46–52
    angular, 139–140
    graphical analysis, 52–54
    motion, 45–46
    motion due to gravity, 51–52
    projectiles launched, 55–57
    relative motion, 54
    uniform circular motion, 57–59
Kinetic energy, 94–95, 122
Kinetic friction, 77

L
Laminar flow, 165
Law(s)
    action-reaction, 69
    conservation, 96–100, 121
    cosine, 35
    Hooke's, 75, 149

235

V

Vectors
    addition of, 33–39
    algebraic considerations, 35–36
    displacement, 45
    geometric considerations, 33–35
    quantity, 32
    resolution, 37–39
    subtraction of, 37
Velocity, 45, 50, 54, 110
Viscosity, 165

W

Weight, 68–69, 74

Work, 91–93
Work-energy theorem, 94–95

Z

Zero acceleration, 50, 71